Springer Theses

Recognizing Outstanding Ph.D. Research

Aims and Scope

The series "Springer Theses" brings together a selection of the very best Ph.D. theses from around the world and across the physical sciences. Nominated and endorsed by two recognized specialists, each published volume has been selected for its scientific excellence and the high impact of its contents for the pertinent field of research. For greater accessibility to non-specialists, the published versions include an extended introduction, as well as a foreword by the student's supervisor explaining the special relevance of the work for the field. As a whole, the series will provide a valuable resource both for newcomers to the research fields described, and for other scientists seeking detailed background information on special questions. Finally, it provides an accredited documentation of the valuable contributions made by today's younger generation of scientists.

Theses are accepted into the series by invited nomination only and must fulfill all of the following criteria

- They must be written in good English.
- The topic should fall within the confines of Chemistry, Physics, Earth Sciences, Engineering and related interdisciplinary fields such as Materials, Nanoscience, Chemical Engineering, Complex Systems and Biophysics.
- The work reported in the thesis must represent a significant scientific advance.
- If the thesis includes previously published material, permission to reproduce this must be gained from the respective copyright holder.
- They must have been examined and passed during the 12 months prior to nomination.
- Each thesis should include a foreword by the supervisor outlining the significance of its content.
- The theses should have a clearly defined structure including an introduction accessible to scientists not expert in that particular field.

More information about this series at http://www.springer.com/series/8790

Arshad Khan

Novel Embedded Metal-mesh Transparent Electrodes

Vacuum-free Fabrication Strategies and Applications in Flexible Electronic Devices

Doctoral Thesis accepted by
The University of Hong Kong,
Pokfulam, Hong Kong

Springer

Author
Dr. Arshad Khan
The University of Hong Kong
Pokfulam, Hong Kong

Supervisor
Prof. Wen-Di Li
Department of Mechanical Engineering
The University of Hong Kong
Pokfulam, Hong Kong

ISSN 2190-5053 ISSN 2190-5061 (electronic)
Springer Theses
ISBN 978-981-15-2917-7 ISBN 978-981-15-2918-4 (eBook)
https://doi.org/10.1007/978-981-15-2918-4

This Springer imprint is published by the registered company Springer Nature Singapore Pte Ltd.
The registered company address is: 152 Beach Road, #21-01/04 Gateway East, Singapore 189721, Singapore

Dedicated to my teachers and my family.

Supervisor's Foreword

It is my pleasure to present Dr. Arshad Khan's Ph.D. thesis work, entitled *Novel Embedded Metal-mesh Transparent Electrodes: Vacuum-free Fabrication Strategies and Applications in Flexible Electronic Devices*, for its publication in Springer Theses Series as an original research work.

This thesis addresses key challenges facing the transparent conductors used in future flexible electronic devices, such as non-smooth surface topography, poor mechanical stability, low fabrication throughput, and high manufacturing cost, by introducing embedded metal-mesh transparent electrodes (EMTE) and vacuum-free fabrication strategies for the novel transparent electrodes. The performance of the device developed in his thesis is among the highest reported values in recent studies.

Dr. Khan's thesis has produced excellent publications in high-quality journals, multiple conference presentations, and a patent application. The device structure and fabrication techniques developed in this thesis have successfully led to real products. Based on this research, a spin-off company named "Flectrode Technology Limited" (www.flectrode.com) has been established through the Technology Start-up Support Scheme for Universities (TSSSU) supported by HKU and Hong Kong SAR government. With more than HKD 40M private investment, the company has set up a pilot production line in Xiamen, China, to commercialize the technology through mass production.

Hong Kong, China
November 2019

Prof. Wen-Di Li

Preface

This study reports a new structure of flexible transparent electrodes featuring a metal mesh fully embedded and mechanically anchored in a flexible substrate, termed an "embedded metal-mesh transparent electrode (EMTE)." In addition, this thesis presents cost-effective and vacuum-free fabrication strategies for this novel transparent electrode and its applications in flexible electronic devices. The embedded nature of EMTE provides a series of advantages, including surface smoothness which is crucial for device fabrication, mechanical stability under high bending stress, strong adhesion to the substrate with excellent flexibility, and favorable resistance against moisture, oxygen, and chemicals. Our novel fabrication techniques are solution-processed and vacuum-free and therefore are potentially suitable for high-throughput, large-volume and low-cost production. In particular, these strategies enable fabrication of high-aspect-ratio (thickness to linewidth) metal meshes, substantially improving conductivity without considerably sacrificing transparency.

Our first vacuum-free fabrication strategy combines lithography, electroplating, and imprint transfer (LEIT) for making EMTEs. We utilized this technique to fabricate various prototype flexible micro-EMTEs and flexible nano-EMTEs with transmittance higher than 90% and sheet resistance below 1 Ω/sq, as well as extremely high figures of merit (ratio of electrical conductivity to optical conductivity) up to 1.5×10^4, which are among the highest reported values in recent studies. Although LEIT is a cost-effective approach for fabrication of EMTEs, it comprises a mandatory lithography step in making each sample which limits its suitability for high-throughput and large-volume industrial production and needs further simplifications.

Therefore, we demonstrate improved techniques based on templated electrodeposition for the fabrication of EMTEs by eliminating the obligatory lithography step from the unit production cycle of LEIT. In these templated electrodeposition and imprint transfer (TEIT) strategies, reusable templates are employed to simplify the fabrication of the EMTEs. Similarly, based on these improved techniques, prototype micro-EMTEs and nano-EMTEs are fabricated on flexible substrates, demonstrating excellent electrical, and optical performances. The mechanical robustness of the reusable templates is also tested by utilizing them for repeated fabrication cycles.

As practical applications, the EMTEs are utilized in flexible bifacial dye-sensitized solar cells (DSSCs) and flexible transparent thin-film heaters (FTTHs). A novel counter electrode (CE) is developed for DSSCs, containing a micro-EMTE with catalytic platinum nanoparticles (PtNPs) in situ electrodeposited only on the surface of the nickel mesh without considerably reducing its optical transparency. The flexible bifacial DSSCs based on this hybrid PtNP-decorated micro-EMTE demonstrate remarkable power conversion efficiencies (PCEs) both under front-side illumination and rear-side illumination. Furthermore based on EMTEs, a FTTH is fabricated and characterized. This device shows a rapid response, requires a low input power density, and operates at ultra-low voltage. These promising results reveal the enormous potential of our EMTEs in the production and commercialization of low-cost and efficient flexible electronic devices.

Pokfulam, Hong Kong — Dr. Arshad Khan

List of Publications

Peer-reviewed Journals

- Arshad Khan, Chuwei Liang, Yu-Ting Huang, Cuiping Zhang, Jingxuan Cai, Shien-Ping Feng, & Wen-Di Li: Template-electrodeposited and imprint-transferred microscale metal-mesh transparent electrodes for flexible and stretchable electronics, ***Advanced Engineering Materials*** (2019), 1900723.
- Arshad Khan, Sangeon Lee, Taehee Jang, Ze Xiong, Cuiping Zhang, Jinyao Tang, L. Jay Guo, & Wen-Di Li: Scalable solution-processed fabrication strategy for high-performance flexible transparent electrodes with embedded metal mesh, ***Journal of Visualized Experiments*** (2017), (124), e56019, https://doi.org/10.3791/56019.
- Arshad Khan[†], Yu-Ting Huang[†], Tsutomu Miyasaka, Masashi Ikegami, Shien-Ping Feng, & Wen-Di Li: Solution-processed transparent nickel-mesh counter electrode with in-situ electrodeposited platinum nanoparticles for full-plastic bifacial dye-sensitized solar cells, ***ACS Applied Materials & Interfaces*** (2017), 9 (9), 8083–8091.
- Arshad Khan, Sangeon Lee, Taehee Jang, Ze Xiong, Cuiping Zhang, Jinyao Tang, L. Jay Guo, & Wen-Di Li: High-performance flexible transparent electrode with an embedded metal mesh fabricated by cost-effective solution process, ***Small*** (2016), 12 (22), 3021–3030.
- Arshad Khan, Shijie Li, Xin Tang, & Wen-Di Li: Nanostructure transfer using cyclic olefin copolymer templates fabricated by thermal nanoimprint lithography, ***Journal of Vacuum Science & Technology B: Microelectronics and Nanometer Structures*** (2014), 32 (6), 06FI02.

Patents

- Arshad Khan & Wen-Di Li: *Transparent Conductive Films with Embedded Metal Grids*, **U.S. Pat. Appl. Publ.** (2016), US2016/0345430 A1; **PCT Int. Appl.** (2016), PCT/CN2016/08110, WO/2016/188308

Conference Presentations

- Arshad Khan, Cuiping Zhang, and Wen-Di Li: Metallic embedded nanomesh as transparent electrode fabricated by template-based electrodeposition for flexible electronic devices, ***2017 MRS Fall Meeting & Exhibit***, Boston, USA, December 2017.
- Arshad Khan, Yu-Ting Huang, Shien-Ping Feng, and Wen-Di Li: Flexible transparent electrode with embedded metal mesh fabricated via template-based electrodeposition for full-plastic bifacial dye-sensitized solar cells, ***The 61st EIPBN Conference***, Orlando, USA, May 2017.

- Arshad Khan, Yu-Ting Huang, Peng Zhai, Shien-Ping Feng, and Wen-Di Li: Cost-effective fabrication of transparent electrode with embedded metal mesh for flexible organic solar cells, ***2016 MRS Spring Meeting & Exhibit***, Phoenix, USA, March 2016.
- Arshad Khan and Wen-Di Li: Cost-effective and solution-processed fabrication for metal mesh-based flexible transparent electrodes, ***The 59th EIPBN Conference***, San Diego, USA, May 2015.
- Arshad Khan, Shijie Li and Wen-Di Li: Fabrication and testing of flexible cyclic olefin copolymer stamps by nanoimprint lithography, ***The 58th EIPBN Conference***, Washington DC, USA, May 2014.

Acknowledgements

By the name of Almighty Allah, the most merciful, the most beneficent. I would like to present my humble gratitude in front of Almighty, Who enabled me to accomplish the dignified cause of education and learning, and I would pray Him that He makes me able to utilize my knowledge and this edification for the betterment of humanity. Ameen!

I am heartily thankful to my supervisor Dr. Wen-Di Li, whose encouragement, guidance, and support from the initial to the final level enabled me to develop an understanding of the subject. I always learn something new whenever I discussed my work with him. I very much appreciate his attitude that if you give it enough thought and creativity, you can engineer your way around any research problem. This book would not have been possible without his keen interest and valuable support.

I extend my thanks to all my laboratory colleagues and coworkers at HKU (Mr. Shijie Li, Dr. Jingxuan Cai, Dr. Zhouyang Zhu, Mrs. Cuiping Zhang, Mr. Tuo Qu, Miss. Chuwei Liang, Miss. Siyi Min, Dr. Ze Xiong, Miss. Yu-Ting Huang) for the fruitful discussions. They have been great resources of encouragement and insight into many topics as well as a great sounding board for my bizarre ideas. The lovely memories I have with these guys will always be cherished. I am also grateful to the staff in the Electron Microscope Unit at HKU for their professional help on characterizations of samples.

Also, I would like to thank our collaborators Dr. Jinyao Tang and Dr. Shien-Ping Feng of HKU, and Prof. L. Jay Guo of the University of Michigan for full-support to access their laboratory facilities for experiments and results analysis.

My deepest gratitude goes to my parents for their spiritual/financial supports. None of my endeavors would be possible without them. Especially, I pay homage to my wife Anjum Bibi and son Mamoon Arshad for their continuous love and understanding.

I am also thankful to our funding organizations, the Research Grants Council of Hong Kong, the National Natural Science Foundation of China, the Science and Technology Innovation Commission of Shenzhen Municipality, the Environmental and Conservation Fund of Hong Kong, and the University of Hong Kong for their financial supports.

Lastly, I offer my regards and blessings to all of those who supported me in any respect during the completion of this project.

Contents

Abbreviations

AFM	Atomic force microscope
AgNW	Silver nanowire
CE	Counter electrode
CNT	Carbon nanotube
COC	Cyclic olefin copolymer
CV	Cyclic voltammograms
DI	Deionized
DMT	Derjaguin–Muller–Toporov
DSSC	Dye-sensitized solar cell
EBL	Electron-beam lithography
EDS	Energy-dispersive spectroscopy
EIS	Electrochemical impedance spectroscopy
EMTE	Embedded metal-mesh transparent electrode
FOM	Figure of merit
FTO	Fluorine-doped tin oxide
FTTH	Flexible transparent thin-film heater
ITO	Indium tin oxide
LEIT	Lithography electroplating imprint transfer
NIL	Nanoimprint lithography
NWL	Nanowire lithography
PCE	Power conversion efficiency
PDMS	Polydimethylsiloxane
PEDOT:PSS	Poly-ethylenedioxythiophene polystyrene sulfonate
PEN	Polyethylene naphthalate
PET	Polyethylene terephthalate
PMMA	Polymethyl methacrylate
PtNP	Platinum nanoparticle
RF	Radio frequency
RIE	Reactive-ion etching
SEM	Scanning electron microscope

TC	Transparent conductor
TCO	Transparent conducting oxide
TEIT	Templated electrodeposition imprint transfer
UV	Ultraviolet
XRD	X-ray diffraction

List of Figures

List of Tables

Chapter 1
Introduction to Transparent Conductors

Abstract Transparent materials are usually insulators while electrically conductive materials are typically opaque at the visible part of the spectrum. Luckily, there is a special class of materials, known as transparent conducting oxides (TCOs), such as indium tin oxide (ITO), fluorine-doped tin oxide (FTO), and zinc oxide etc. have a blend of properties that make them both electrically conductive and highly transparent in the visible range. Among these oxides films, ITO has the industry standard role and has shown its potential as transparent conductor (TC) in variety of electronic devices such as light-emitting diodes (Wu et al. in ACS Nano 4:43, 2010 [1]), solar cells (Yu et al. in Adv Mater 23:4453, 2011 [2]), displays (Blake et al. in Nano Lett 8:1704, 2008 [3]), touchscreens (Wang et al. in Adv Mater 24:2874, 2012 [4]), and smart windows (Deb et al. in Electr Acta 46:2125, 2001 [5]). In addition to low sheet resistance and high optical transparency needed for conventional TCs, emerging flexible electronic devices also demand excellent flexibility (Hecht et al. in Adv Mater 23:1482, 2011 [6]; Ellmer in Nat Photon 6:809, 2012 [7]) in next-generation TCs. Although ITO-based TCs have exhibited desirable performance such as high conductivity, high transmittance in visible region and decent environmental stability, many concerns, including film brittleness (Cairns et al. in Appl Phys Lett 76:1425, 2000 [8]), low infrared transmittance (Bel Hadj Tahar et al. in J Appl Phys 83:2631, 1998 [9]), low abundance (Kumar and Zhou in ACS Nano 4:11, 2010 [10]), and inability of high temperature annealing on plastic substrates, limit its suitability for use in next-generation flexible electronic devices. Therefore, novel TCs other than TCOs must be developed to address the challenges in these emerging electronic devices. In order to provide the readers a general background, brief overview of TCs is presented here in this chapter.

1.1 Organic Alternatives to Indium Tin Oxide

Similar to other topics in materials science, the most common approach to sub-categorized ITO alternatives TCs is to divide these into organic and inorganic. Organic alternatives are: carbon nanotube networks (CNTs), graphene, and some other conducting polymers.

A. Khan, *Novel Embedded Metal-mesh Transparent Electrodes*, Springer Theses,
https://doi.org/10.1007/978-981-15-2918-4_1

1.1.1 Carbon Nanotube Networks

CNTs are potential alternative material for flexible TCs due to its high mechanical robustness as compared to ITO films. CNTs are one of the hardest materials acknowledged; along with excellent electric properties and many other unique features [11–14]. Because of these unique electronic and mechanical characteristics, CNTs have been successfully utilized over the years as TC in flexible electronic devices [14–16]. Several techniques such as transfer printing [17], vacuum filtration [18], and spin coating [19] have been established to fabricate CNTs based TCs with the required optical and electrical properties for organic solar cells, flexible displays, light emitting diodes, and touch screens. Although CNTs have promising properties, for instance it has high optical transparency and excellent mechanical flexibility, high-performance large-area flexible devices are difficult to achieve using CNTs because of its lower conductivity. The sheet resistance of CNTs is much higher than that of typical ITO-based TCs. As a result, its performance remains poor and is not suitable for commercial use in large-area flexible electronic devices.

1.1.2 Graphene

Graphene has been recommended as an alternate TCs material because of its unique properties [20, 21]. It is almost 200 times stronger than steel, efficiently conducts heat and electricity, and is almost fully transparent [20]. Different methodologies are developed to fabricate graphene films on flexible substrates, including chemical vapor deposition growth on catalyst surfaces, reduction of graphene oxide, graphite intercalation and graphene flake deposition [22–24]. In recent years remarkable progress has been made to improve the opto-electrical performance of the graphene based TCs. Large-area graphene on copper-catalyst (30 inches diagonal) has been fabricated, and successfully transferred to flexible polymer substrates by contact transfer method, demonstrated remarkable electrical and optical performance of 30 Ω/sq and 90% transparency respectively [25]. However, this outstanding performance is difficult to reproduce, the typical reported sheet resistance and optical transmittance values for graphene based TCs are still hundreds of Ω/sq and around 80% respectively [26]. Ideally, graphene has enormous potential and is holding the promise of being the ultimate TC, but in reality it is extremely difficult and presently costly to fabricate its uniform films. Furthermore, due to crystallographic defects, folds, and wrinkles it's electrical and optical performances degrade rapidly [25, 27].

1.1.3 Other Conducting Polymers

Charge transfers in intrinsically conducting polymers by long chains of delocalized bonds (aromatic cycles, carbon double bond). However, the electrical conductivity of this class of materials is very low as compare to that of ITO-based TCs due to the shortage of free carriers. In addition to poor conductance, intrinsically conducting polymers are also known for visible absorptive resonances, which makes most of them opaque in visible part of the spectrum [28, 29]. However, some of the intrinsically transparent conducting polymers are converted into charge transfer polymers by adding conductive dopants into its repeating units. One of the typical example of charge transfer polymers is poly (3,4-ethylenedioxythiophene) polystyrene sulfonate (PEDOT:PSS). In PEDOT: PSS, PEDOT plays the role of intrinsically conducting polymer, while the PSS acts as dopant which enhances the electrical conductivity by efficiently concentrating the negative carriers. Furthermore, PEDOT: PSS has no visible absorptive resonances, and is therefore commonly used as TC at small-scale in the laboratory. However, several issues such as water solubility, molecular instability, and other degradations have retained PEDOT: PSS from large-scale industrial applications [30, 31].

1.2 Inorganic Alternatives to Indium Tin Oxide

Inorganic TCs are considered the leading candidates for substituting ITO-based TCs. Inorganic alternate TCs are based on transparent thin metal films, random metal nanowire networks, and regular metal meshes.

1.2.1 Transparent Thin Metal Films

In most electronic devices, thick metallic films (tens to hundreds of nm) are utilized as back-electrodes (cathodes). However, thin (few nm) metallic films can also be used as transparent front-electrodes (anodes). These films are very thin as compared to the visible wavelength of light and therefore are optically transparent. Many metals with various work functions such as gold [32], nickel [33], silver [34–36], and platinum [37] etc. have been successfully used as anodes in flexible electronic devices. The electrical conductivity and optical transmittance of the thin metal film based TCs can be optimized for the required applications by keeping its thickness and quality uniform over the whole area of the substrate. But, the fabrication of ultra-thin and uniform metal films over large substrate is challenging and therefore significant progresses in the fabrication techniques must be done before these thin metal films can be considered as potential alternatives for the traditional ITO-based TCs.

1.2.2 Metal Nanowire Networks

One of the promising categories of inorganic TCs is made from random metal nanowire networks [38, 39] which have showed great potential in electrical conductivity, optical transparency, and mechanical flexibility. In particular, TCs made by the percolation of random silver nanowires (AgNWs) got strong interest due to the fact that they can be distributed in ink and printed in low-cost solution-based processing, which is desirable when forming flexible electronic devices on large-area flexible substrates [40]. However, AgNW networks suffer from several problems such as difficulty in attaining uniform nanowire spreading over the whole substrate and the delamination of the nanowires from a substrate [7]. Also, the randomly dispersed nanowire network cannot be used readily; extra processes are usually required to eliminate the capping polymer around the nanowire and reduce junction resistance, either by selective welding [41], bulk heating [42], or other chemical adjustments [43, 44].

1.2.3 Regular Metal Meshes

Compared with AgNWs-based TCs, regular metal-mesh TCs seem more promising as their electrical conductivity and optical transmittance can be tuned in a wide range by adjusting the line spacing, line width, and metal thickness [45]. Furthermore, unlike the limited material choices for nanowires, various metallic materials can be used in regular metal mesh-based TCs to achieve desirable chemical properties and different work functions for required applications [46]. Yet, widespread adoption of metal-mesh based TCs has been limited by several key issues such as expensive vacuum-based metal deposition from the vapor phase, non-flat surface topography and weak adhesion between the metal-mesh and flexible substrate. Some current research has been building progress to resolve these issues. Embedding the silver colloidal nanoparticles in embossed trenches on a plastic film [47] has been utilized for commercial fabrication, however limited metals can straightforwardly form nanoparticles and the decreased conductivity of annealed nanoparticles also damage their performance. Fabrication of solution-grown AgNWs by electroless plating was used to make ordered arrays [48]. But, this technique can be utilized for very few metals and its scalability for fabrication is too limited. The above-mentioned challenges demand for novel metal-mesh TCs structures as well as better and scalable fabrication approaches for their manufacturing. This thesis particularly focuses to address these key challenges which limit widespread applications of metal-mesh TCs in flexible electronics industry.

1.3 Thesis Organization

This dissertation discusses in detail the embedded metal-mesh transparent electrodes with emphasis on its unique advantageous structure, and cost-effective solution-based fabrication approaches for this novel transparent electrodes. In order to provide the readers a general background, brief overview of TCs is presented in Chap. 1 of this thesis. Chapter 2 generally introduces the EMTE structure and summarizes the methodologies used in key steps of its fabrication. Chapter 3 fully describes one of our novel LEIT fabrication approach for making micro-EMTEs. As a demonstration, various prototype copper micro-EMTEs with transmittance higher than 90% and sheet resistance below 1 Ω/sq, as well as extremely high figures of merit up to 1.5×10^4 are fabricated on flexible films. Similarly, Chap. 4 concentrates on our improved TEIT fabrication methodology for making micro-EMTEs. Based on this TEIT technique, prototype micro-EMTEs are fabricated on flexible substrates, demonstrating excellent electrical and optical performances. Chapter 5 presents the dimensional scalability of LEIT and TEIT fabrication strategies by utilizing them for production of nano-EMTEs. The applications of EMTEs in flexible electronic devices, i.e. flexible bifacial DSSCs and FTTHs, are comprehensively discussed in Chap. 6. Finally, a brief summary of the key accomplishments and future recommendations are presented in Chap. 7.

References

1. Wu J, Agrawal M, Becerril HA, Bao Z, Liu Z, Chen Y, Peumans P (2010) Organic light-emitting diodes on solution-processed graphene transparent electrodes. ACS Nano 4(1):43–48. https://doi.org/10.1021/nn900728d
2. Yu Z, Li L, Zhang Q, Hu W, Pei Q (2011) Silver Nanowire-polymer composite electrodes for efficient polymer solar cells. Adv Mater 23(38):4453–4457. https://doi.org/10.1002/adma.201101992
3. Blake P, Brimicombe PD, Nair RR, Booth TJ, Jiang D, Schedin F, Ponomarenko LA, Morozov SV, Gleeson HF, Hill EW, Geim AK, Novoselov KS (2008) Graphene-based liquid crystal device. Nano Lett 8(6):1704–1708. https://doi.org/10.1021/nl080649i
4. Wang J, Liang M, Fang Y, Qiu T, Zhang J, Zhi L (2012) Rod-Coating: towards large-area fabrication of uniform reduced graphene oxide films for flexible touch screens. Adv Mater 24(21):2874–2878. https://doi.org/10.1002/adma.201200055
5. deb, S. K., Lee, S., Edwin Tracy, C., Roland Pitts, J., Gregg, B. A., Branz, H. M. (2001) Stand-alone photovoltaic-powered electrochromic smart window. Electrochim Acta 46(13):2125–2130. https://doi.org/10.1016/S0013-4686(01)00390-5
6. Hecht DS, Hu L, Irvin G (2011) Emerging transparent electrodes based on thin films of carbon nanotubes, graphene, and metallic nanostructures. Adv Mater 23(13):1482–1513. https://doi.org/10.1002/adma.201003188
7. Ellmer K (2012) Past achievements and future challenges in the development of optically transparent electrodes. Nat Photon 6(12):809–817
8. Cairns DR, Witte RP, Sparacin DK, Sachsman SM, Paine DC, Crawford GP, Newton RR (2000) Strain-dependent electrical resistance of tin-doped indium oxide on polymer substrates. Appl Phys Lett 76(11):1425–1427. https://doi.org/10.1063/1.126052

9. Tahar Bel Hadj R, Ban T, Ohya Y, Takahashi Y (1998) Tin doped indium oxide thin films: electrical properties. J Appl Phys 83(5):2631–2645. https://doi.org/10.1063/1.367025
10. Kumar A, Zhou C (2010) The race to replace tin-doped indium oxide: which material will win?. ACS Nano 4(1):11–14. https://doi.org/10.1021/nn901903b
11. Baughman RH, Zakhidov AA, de Heer WA (2002) Carbon nanotubes–the route toward applications. Sci 297(5582):787–792. https://doi.org/10.1126/science.1060928
12. Iijima S (1991) Helical microtubules of graphitic carbon. Nat 354(6348):56–58
13. Tasis D, Tagmatarchis N, Bianco A, Prato M (2006) Chemistry of carbon nanotubes. Chem 106(3):1105–1136. https://doi.org/10.1021/cr050569o
14. Wu Z, Chen Z, Du X, Logan JM, Sippel J, Nikolou M, Kamaras K, Reynolds JR, Tanner DB, Hebard AF, Rinzler AG (2004) Transparent, conductive carbon nanotube films. Sci 305(5688):1273–1276. https://doi.org/10.1126/science.1101243
15. Lagemaat Jvd, Barnes TM, Rumbles G, Shaheen SE, Coutts TJ, Weeks C, Levitsky I, Peltola J, Glatkowski P (2006) Organic solar cells with carbon nanotubes replacing In2O3: Sn as the transparent electrode. Appl Phys Lett 88(23):233503. https://doi.org/10.1063/1.2210081
16. Rowell MW, Topinka MA, McGehee MD, Prall H-J, Dennler G, Sariciftci NS, Hu L, Gruner G (2006) Organic solar cells with carbon nanotube network electrodes. Appl Phys Lett 88(23):233506. https://doi.org/10.1063/1.2209887
17. Pint CL, Xu Y-Q, Moghazy S, Cherukuri T, Alvarez NT, Haroz EH, Mahzooni S, Doorn SK, Kono J, Pasquali M, Hauge RH (2010) Dry contact transfer printing of aligned carbon nanotube patterns and characterization of their optical properties for diameter distribution and alignment. ACS Nano 4(2):1131–1145. https://doi.org/10.1021/nn9013356
18. Benjamin K, Balaji P (2014) Vacuum filtration based formation of liquid crystal films of semiconducting carbon nanotubes and high performance transistor devices. Nanotechnol 25(17):175201
19. Jo JW, Jung JW, Lee JU, Jo WH (2010) Fabrication of highly conductive and transparent thin films from single-walled carbon nanotubes using a new non-ionic surfactant via spin coating. ACS Nano 4(9):5382–5388. https://doi.org/10.1021/nn1009837
20. Mattevi C, Eda G, Agnoli S, Miller S, Mkhoyan KA, Celik O, Mastrogiovanni D, Granozzi G, Garfunkel E, Chhowalla M (2009) Evolution of electrical, chemical, and structural properties of transparent and conducting chemically derived graphene thin films. Adv Funct Mater 19(16):2577–2583. https://doi.org/10.1002/adfm.200900166
21. Wassei JK, Kaner RB (2010) Graphene, a promising transparent conductor. Mater Today 13(3):52–59. https://doi.org/10.1016/S1369-7021(10)70034-1
22. Park S, Lee K-S, Bozoklu G, Cai W, Nguyen ST, Ruoff RS (2008) Graphene oxide papers modified by divalent ions—enhancing mechanical properties via chemical cross-linking. ACS Nano 2(3):572–578. https://doi.org/10.1021/nn700349a
23. Reina A, Jia X, Ho J, Nezich D, Son H, Bulovic V, Dresselhaus MS, Kong J (2009) Large Area, few-layer graphene films on arbitrary substrates by chemical vapor deposition. Nano Lett 9(1):30–35. https://doi.org/10.1021/nl801827v
24. De S, King PJ, Lotya M, O'Neill A, Doherty EM, Hernandez Y, Duesberg GS, Coleman JN (2010) Flexible, transparent, conducting films of randomly stacked graphene from surfactant-stabilized. Oxide-Free graphene dispersions. Small 6(3):458–464. https://doi.org/10.1002/smll.200901162
25. Bae S, Kim H, Lee Y, Xu X, Park J-S, Zheng Y, Balakrishnan J, Lei T, Ri Kim H, Song YI, Kim Y-J, Kim KS, Ozyilmaz B, Ahn J-H, Hong BH, Iijima S (2010) Roll-to-roll production of 30-inch graphene films for transparent electrodes. Nat Nano 5(8):574–578
26. Biswas C, Lee YH (2011) Graphene versus carbon nanotubes in electronic devices. Adv Funct Mater 21(20):3806–3826. https://doi.org/10.1002/adfm.201101241
27. Bae S-Y, Jeon I-Y, Yang J, Park N, Shin HS, Park S, Ruoff RS, Dai L, Baek J-B (2011) Large-Area graphene films by simple solution casting of edge-selectively functionalized graphite. ACS Nano 5(6):4974–4980. https://doi.org/10.1021/nn201072m
28. Patil AO, Heeger AJ, Wudl F (1988) Optical properties of conducting polymers. Chem Rev 88(1):183–200. https://doi.org/10.1021/cr00083a009

29. Stenger-Smith JD (1998) Intrinsically electrically conducting polymers. Synthesis, characterization, and their applications. Prog Polym Sci 23(1):57–79. https://doi.org/10.1016/S0079-6700(97)00024-5
30. Elschner A, Kirchmeyer S, Lovenich W, Merker U, Reuter K (2010) PEDOT: principles and applications of an intrinsically conductive polymer. CRC Press,: FL
31. Lipomi DJ, Lee JA, Vosgueritchian M, Tee BCK, Bolander JA, Bao Z (2012) Electronic properties of transparent conductive films of PEDOT:PSS on stretchable substrates. Chem Mater 24(2):373–382. https://doi.org/10.1021/cm203216m
32. Lin K, Kumar RS, Peng C, Lu S, Soo-Jin C, Burden AP (2005) Au-ITO anode for efficient polymer light-emitting device operation. IEEE Photonic Tech L 17(3):543–545. https://doi.org/10.1109/LPT.2004.841031
33. Lee CJ, Pode RB, Moon DG, Han JI (2004) Realization of an efficient top emission organic light-emitting device with novel electrodes. Thin Solid Films 467(1):201–208. https://doi.org/10.1016/j.tsf.2004.04.016
34. Zhijun W, Shufen C, Huishan Y, Yi Z, Jingying H, Shiyong L (2004) Top-emitting organic light-emitting devices based on silicon substrate using Ag electrode. Semicond Sci Technol 19(9):1138
35. Peng H, Zhu X, Sun J, Xie Z, Xie S, Wong M, Kwok H-S (2005) Efficient organic light-emitting diode using semitransparent silver as anode. Appl Phys Lett 87(17):173505. https://doi.org/10.1063/1.2115076
36. Gu D, Zhang C, Wu Y-K, Guo LJ (2014) Ultrasmooth and thermally stable silver-based thin films with subnanometer roughness by aluminum doping. ACS Nano 8(10):10343–10351. https://doi.org/10.1021/nn503577c
37. Chengfeng Qiu HPHCZXMWHSK (2004) Top-emitting OLED using praseodymium oxide coated platinum as hole injectors. IEEE Trans Electron Devices 51:1207
38. De S, Higgins TM, Lyons PE, Doherty EM, Nirmalraj PN, Blau WJ, Boland JJ, Coleman JN (2009) Silver nanowire networks as flexible, transparent, conducting films: extremely high dc to optical conductivity ratios. ACS Nano 3(7):1767–1774. https://doi.org/10.1021/nn900348c
39. van de Groep J, Spinelli P, Polman A (2012) Transparent conducting silver nanowire networks. Nano Lett 12(6):3138–3144. https://doi.org/10.1021/nl301045a
40. Jiu J, Nogi M, Sugahara T, Tokuno T, Araki T, Komoda N, Suganuma K, Uchida H, Shinozaki K (2012) Strongly adhesive and flexible transparent silver nanowire conductive films fabricated with a high-intensity pulsed light technique. J Mater Chem 22(44):23561–23567. https://doi.org/10.1039/C2JM35545K
41. Garnett EC, Cai W, Cha JJ, Mahmood F, Connor ST, Greyson Christoforo M, Cui Y, McGehee MD, Brongersma ML (2012) Self-limited plasmonic welding of silver nanowire junctions. Nat Mater 11(3):241–249
42. Hu L, Kim HS, Lee J-Y, Peumans P, Cui Y (2010) Scalable coating and properties of transparent, flexible. Silver nanowire electrodes. ACS Nano 4(5):2955–2963. https://doi.org/10.1021/nn1005232
43. Zhu R, Chung C-H, Cha KC, Yang W, Zheng YB, Zhou H, Song T-B, Chen C-C, Weiss PS, Li G, Yang Y (2011) Fused silver nanowires with metal oxide nanoparticles and organic polymers for highly transparent conductors. ACS Nano 5(12):9877–9882. https://doi.org/10.1021/nn203576v
44. Chung C-H, Song T-B, Bob B, Zhu R, Yang Y (2012) Solution-processed flexible transparent conductors composed of silver nanowire networks embedded in indium tin oxide nanoparticle matrices. Nano Res 5(11):805–814. https://doi.org/10.1007/s12274-012-0264-8
45. Kim H-J, Lee S-H, Lee J, Lee E-S, Choi J-H, Jung J-H, Jung J-Y, Choi*, D.-G. (2014) High-Durable AgNi nanomesh film for a transparent conducting electrode. Small 10(18):3767–3774. https://doi.org/10.1002/smll.201400911
46. Wu H, Kong D, Ruan Z, Hsu P-C, Wang S, Yu Z, Carney TJ, Hu L, Fan S, Cui Y (2013) A transparent electrode based on a metal nanotrough network. Nat Nano 8(6):421–425
47. Zhou L, Xiang H-Y, Shen S, Li Y-Q, Chen J-D, Xie H-J, Goldthorpe IA, Chen L-S, Lee S-T, Tang J-X (2014) High-Performance flexible organic light-emitting diodes using embedded

silver network transparent electrodes. ACS Nano 8(12):12796–12805. https://doi.org/10.1021/nn506034g
48. Sciacca B, van de Groep J, Polman A, Garnett EC (2015) Solution-Grown silver nanowire ordered arrays as transparent electrodes. Adv Mater 28(5):905–909. https://doi.org/10.1002/adma.201504045

Chapter 2
Introduction to Vacuum-free Fabrication Strategies for Embedded Metal-mesh Transparent Electrodes

Abstract This chapter introduces EMTE structure and summarizes the key steps in its fabrication. These solution-based fabrication strategies comprises of three major stages i.e. mesh-template patterning on conductive substrate, metal deposition into the mesh-template, and metal-mesh transfer to flexible substrates. We accomplished each of the said steps by utilizing several vacuum-free approaches. For mesh patterning, the methods include photolithography (with and without a stepper), EBL, and NWL while for metal deposition, electrodeposition and electro-less deposition were employed. Similarly, inspired from the thermal NIL and UV-NIL on plastic COC films, imprint transfer methods i.e. thermal imprint transfer and UV-imprint transfer are developed and applied for transferring the metal-meshes onto plastic COC films in embedded form.

2.1 Current Challenges with Metal-mesh Transparent Electrodes

Transparent conductors based on metal-mesh are leading candidates for flexible electronic devices due to their regular structure, excellent electrical conductivity, and flexibility. However, the extensive adoption of metal-mesh TCs has been held up by several problems. First, making metal-mesh TCs often comprises the physical deposition of metal materials from the vapor phase, which involves costly vacuum-based processing [1–4]. Second, a thick layer of metal mesh on the substrate, as essential to attain appropriately high conductivity in numerous applications, might easily cause electrical short circuiting [5–8]. Third, the weak adhesion between the substrate surface and metal mesh results in poor consistency, especially in highly flexible electronic devices [9, 10]. We addressed most of the aforementioned issues by introducing the EMTE structure featuring a metal-mesh fully embedded and mechanically anchored in a flexible substrate.

A. Khan, *Novel Embedded Metal-mesh Transparent Electrodes*, Springer Theses,
https://doi.org/10.1007/978-981-15-2918-4_2

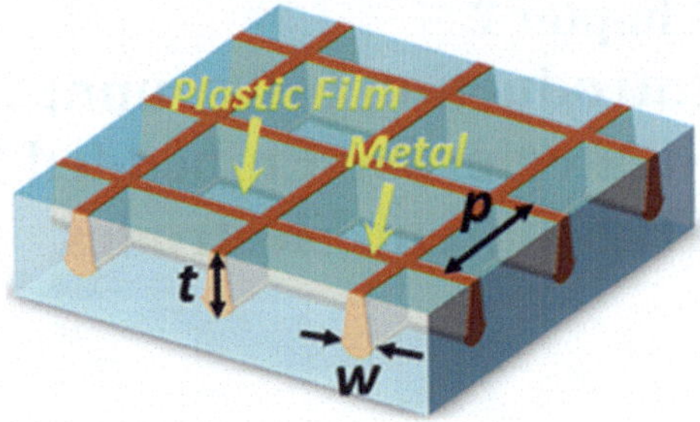

Fig. 2.1 Schematic illustrations of the EMTE structure with a metal-mesh fully embedded in a transparent flexible plastic film

2.2 Structure of EMTEs

The structure of the EMTEs allows the use of thick metal mesh for high conductivity without sacrificing surface smoothness. As displayed in Fig. 2.1, the EMTE comprises a metal-mesh embedded in a flexible substrate. The top surface of the metal-mesh is on the same level as the substrate surface, featuring an overall smooth surface for following device fabrication. The lower part of the metal-mesh has a wider lateral dimension than the upper part does and thus can be mechanically anchored in the substrate to provide mechanical stability. The dimensions of the metal mesh, namely the pitch p, linewidth w, and mesh thickness t, determine the EMTE's electrical conductivity and optical transmittance. Because these dimensional parameters can be tailored in our fabrication processes, the EMTEs have a wide range of artificially designed properties to meet different device requirements and be constructed using a wide selection of metals and thermoplastic substrate materials.

2.3 Key Steps in Fabrication of EMTEs

Fabrication of EMTEs involves three major steps i.e. mesh-template patterning on conductive substrate, metal deposition into the mesh-template, and metal-mesh transfer to flexible substrates. Each of the aforementioned steps is executed by employing numerous solution-based approaches. This section summarized the methodologies used for cost-effective fabrication of the micro-EMTEs and nano-EMTEs.

2.3.1 Mesh-Template Patterning

Mesh-template patterning is a crucial step in EMTEs fabrication as it defines the important dimensional parameters of metal mesh such as pitch, linewidth and cross-sectional shape. In this work, several patterning methodologies including photolithography (with and without a stepper), electron beam lithography (EBL), and nanowire lithography (NWL) are utilized for mesh-template patterning.

2.3.1.1 Photolithography

Photolithography is a procedure used in micro and nano fabrication to pattern parts of thin film on large-area substrates. It utilizes light to transfer the geometric patterns from a photomask to a light-sensitive photoresist on the substrate. The schematic of typical positive-tone photolithography is shown in Fig. 2.2.

In fabrication of micro-EMTEs, we extensively utilized photolithography in both LEIT and TEIT fabrication strategies. During LEIT process, first FTO glass substrates were cleaned and the positive photoresist (AZ 1500) was then spin coated at different speeds onto it. The photoresist was then baked on a hotplate. After allowing it to return to room temperature, the photoresist was exposed to UV light. The photoresist was then developed by immersing it in developer solution. It was finally rinsed in deionized (DI) water and blown dry with compressed air before inspection using light microscope, atomic force microscope (AFM), and scanning electron microscope (SEM) to characterize the fidelity of the features. Square mesh patterns with 2 μm

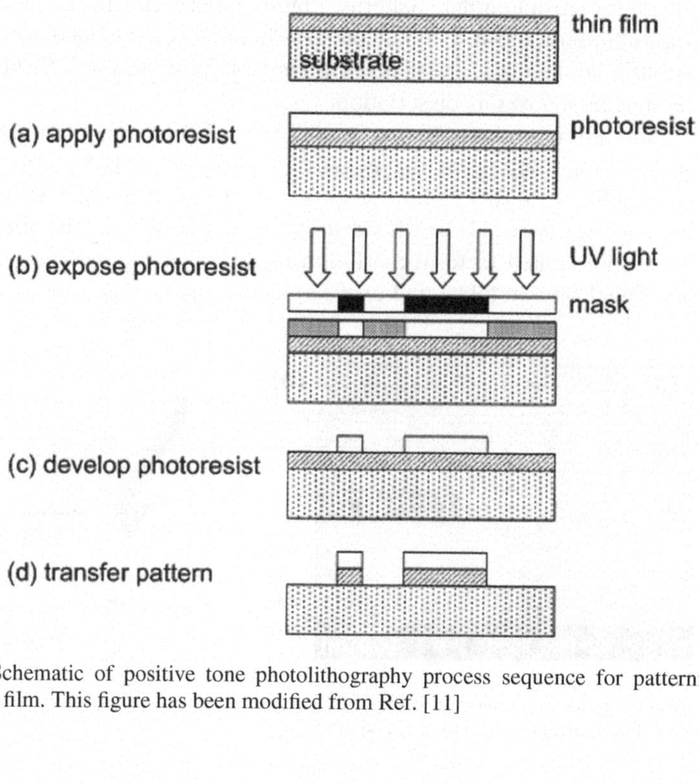

Fig. 2.2 Schematic of positive tone photolithography process sequence for patterning a thin photoresist film. This figure has been modified from Ref. [11]

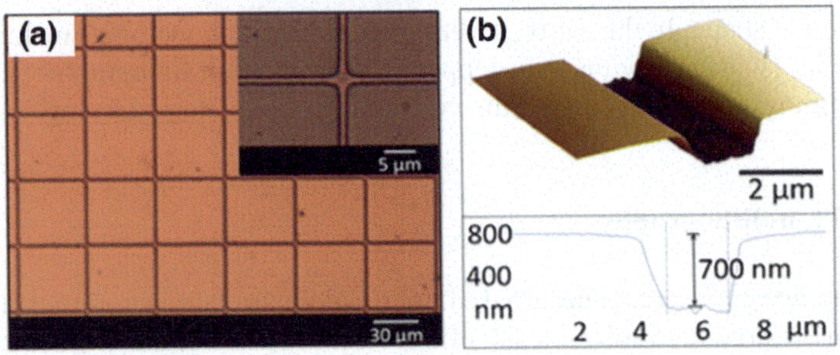

Fig. 2.3 Optical microscope (**a**) and atomic force microscope (**b**) images of micro-mesh pattern in photoresist film

linewidth and various pitches are made in photoresist (AZ 1500) films of different thicknesses. As a typical example, Fig. 2.3 displays the light microscope and AFM images of the 50 µm pitch square micro-mesh pattern in 700 nm thick photoresist film. In similar fashion, other micro-mesh patterns are made in photoresist films which are discussed in-depth in Chap. 3 of this dissertation.

Beside micro-EMTE fabrication by LEIT strategy, photolithography with a stepper is also utilized in nano-EMTEs mesh-template patterning. After spin-coating and baking the positive photoresist on ITO glass substrate, an i-line stepper was used for patterning nano-mesh template patterns. Figure 2.4 presents the SEM and AFM images of the submicron mesh-template patterns in photoresist coated on ITO glass substrate. Additional details of the nano-EMTEs fabrication through LEIT strategy are presented in Chap. 5 of this dissertation.

In TEIT fabrication strategy of micro-EMTEs, SU-8 reusable templates were made by photolithography. First the negative photoresist (SU-8 2000, contains 75% solids) is diluted to 10% solid contents concentration using a thinner solution (cyclopentanone). Next, ITO glass substrates were cleaned and the diluted SU-8 2000 was then spin coated to form a 500 nm thick uniform film. The photoresist was then pre-baked on a hotplate and exposed to UV light. After, post-baking and

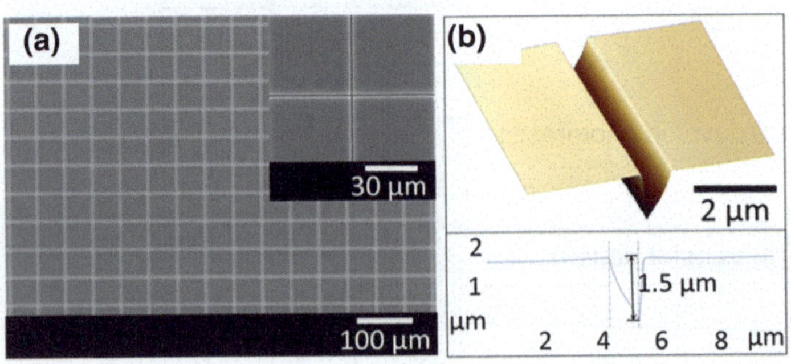

Fig. 2.4 Morphological characterization by SEM (**a**) and AFM (**b**) of a prototype nano-mesh template exposed in photoresist film by a UV stepper

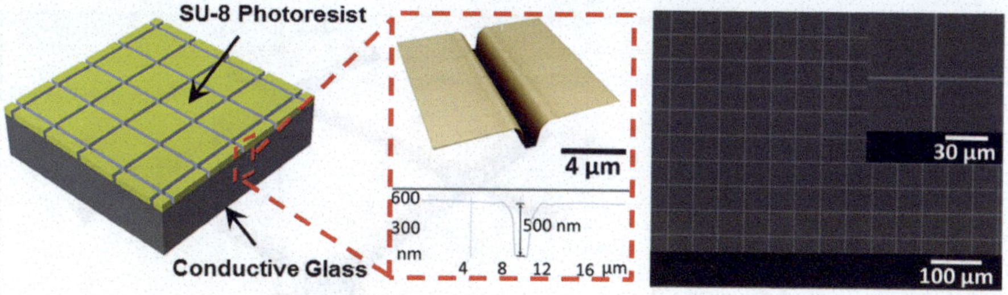

Fig. 2.5 Schematic illustration and morphological characterizations (AFM (left) and SEM (right)) of micro-mesh patterns formed in SU-8 photoresist coated on ITO glass substrate by photolithography

development, the templates were hard-baked at higher temperature so that the SU-8 film firmly adhere with the ITO glass. Figure 2.5 presents the schematic illustration and morphological characterizations (AFM and SEM images) of the SU-8 based reusable template. Further details of the TEIT fabrication and characterizations are documented in Chap. 4 of this dissertation.

2.3.1.2 Electron Beam Lithography

EBL is the practice of scanning a focused electron beam to write custom contours on a substrate coated with electron-sensitive resist as described in Fig. 2.6. The objective of EBL is to make very small shapes in the resist film that can afterwards be transferred to the substrate materials, usually by etching [12]. The primary advantage of EBL is that it can directly sketch custom patterns with very high resolution.

We prepared nano-mesh patterns in polymethyl methacrylate (PMMA) film by EBL to validate the dimensional scalability of LEIT fabrication strategy. Figure 2.7 presents the AFM image of trenches created in PMMA film by e-beam lithography. Detailed information of this pattering process is presented in Chap. 5.

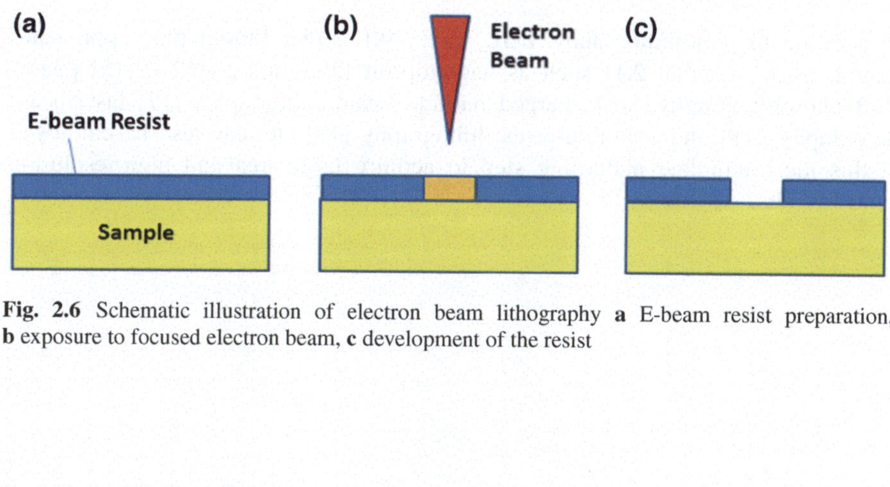

Fig. 2.6 Schematic illustration of electron beam lithography **a** E-beam resist preparation, **b** exposure to focused electron beam, **c** development of the resist

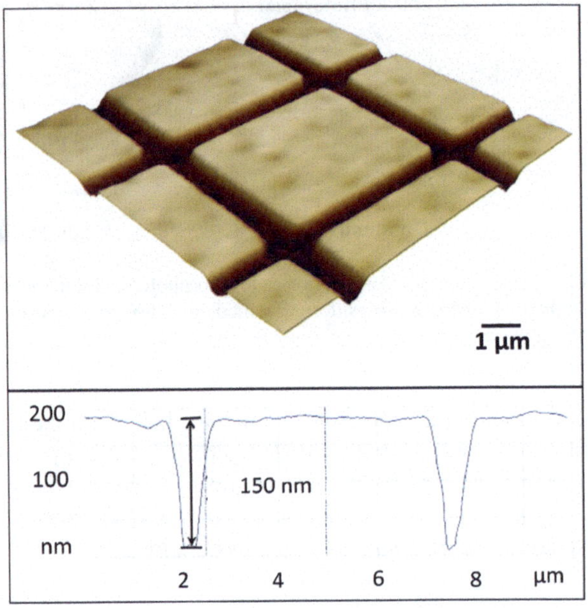

Fig. 2.7 AFM image of the nano-mesh patterns in PMMA film made by EBL

2.3.1.3 Nanowire Lithography

NWL is relative new lithographic approach which utilizes nanowires, arranged by chemical procedures, as etch masks to transfer their 1-dimentional shape to the substrate underneath [13, 14]. Here in this dissertation, we used this technique to pattern nanomesh-template in silicon dioxide (SiO_2) film. This template is then employed in TEIT fabrication of nano-EMTEs. Further details of the methodology of NWL and characterizations are presented in Chap. 5 of this dissertation.

2.3.1.4 Other Proposed Lithographic Approaches

In addition to photolithography, EBL, and NWL, other lithographic approaches (summarized in Table 2.1) such as nanoimprint lithography (NIL), [15] phase-shift photolithography, [16] charged particle beam lithography, [17] laser-beam lithography, [18] and scanning-probe lithography [19] etc. can also be employed in this mesh-template patterning step to acquire large-area and high-resolution nano-EMTEs.

Table 2.1 Summary of the proposed lithographic approaches for mesh-template patterning

Lithography technique	Resolution	Throughput	Mask/Mold	Cost
Photolithography	Few μm	High	Yes	Low
Stepper photolithography	Few hundreds of nm	Medium	Yes	High
Electron-beam lithography	Few tens of nm	Low	No	High
Nanowire lithography	Few tens of nm	High	No	Low
Nanoimprint lithography	Few tens of nm	High	Yes	Low
Phase-shift photolithography	Few tens of nm	High	Yes	High
Particle beam lithography	Few tens of nm	Low	No	High
Laser-beam lithography	Few tens of nm	High	No	High
Scanning-probe lithography	Few tens of nm	Low	No	High

2.3.2 *Metal Deposition*

Metal deposition is a critical step in the fabrication of EMTEs as it affects the morphology of the metal-mesh and the final performance. Our novel fabrication approaches replaces traditional vacuum-based metal deposition with solution-based metal deposition processes and therefore are suitable for high-throughput, large-volume and low-cost production. For metal deposition in EMTEs fabrication, we utilized two solution-based processes namely electrodeposition and electroless-deposition. This section summarizes the details of both processes.

2.3.2.1 Electrodeposition

Electrodeposition has been around for a very long time now; no one knows for sure when and where it instigated [20]. It is a process that uses electric current to reduce dissolved metal cations so that they form a thin uniform metal film on cathode. It is advantageous due to its material versatility and suitability for high-throughput, and low-cost productions. We employed electrodeposition in LEIT and TEIT processes to make both micro-EMTEs and nano-EMTEs by depositing metal inside the lithographically defined trenches as displayed in Fig. 2.8. In this sample electrodeposition step in TEIT process, a constant current is applied for specific time to achieve the desired thickness of deposited metal.

2.3.2.2 Electroless Deposition

Electroless deposition is a classical solution-based method for metal deposition that has been in use for centuries. Despite relatively slow improvements through the centuries, substantial systematic results have been accomplished in the last few decades that contributed to wide-range applications in numerous industries [21]. We also

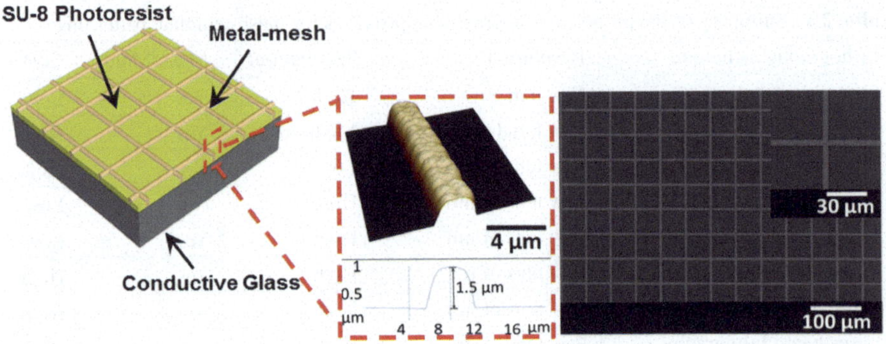

Fig. 2.8 Schematic illustration and morphological characterizations (AFM (left) and SEM (right)) of electrodeposited metal inside the trenches of the SU-8 template

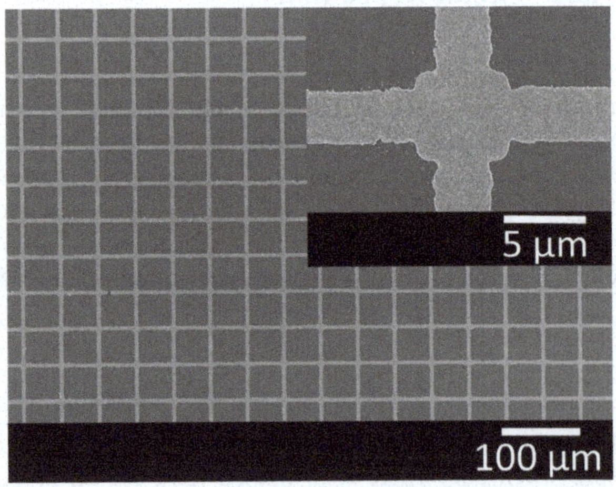

Fig. 2.9 SEM images of copper metal-mesh fabricated by electroless deposition

utilized electroless deposition in fabrication of EMTEs. Figure 2.9 shows the SEM images of the copper metal-mesh TC fabricated by electroless deposition.

2.3.3 *Mesh Transfer to Flexible Substrate*

Metal-mesh transfer and embedding is key step in successful realization of EMTEs. We developed the imprint transfer processes for transferring the metal-mesh to flexible substrates in embedded form after getting motivation from NIL on flexible cyclic olefin copolymer (COC) films [22]. Featureless COC films were first patterned by a thermal NIL using silicon molds with gratings of various periods. Morphology of COC gratings imprinted at different processing parameters was characterized and the grating transfer fidelity was systematically investigated. Based on the excellent thermoplastic properties of COC films towards nanopattern transfer fidelity, we demonstrated the thermal imprint transfer of micro-meshes from conductive glass substrates to these films in embedded form for EMTEs fabrication. Moreover, we also presented the transfer of nano-meshes onto ultraviolet (UV)-curable epoxy coated COC films by UV imprint transfer process. This section presents the summary of both thermal and UV-imprint transfers to plastic films.

2.3.3.1 Motivation from Thermal NIL on Plastic COC Films for Thermal Imprint Transfer of Metal-Meshes to Plastic COC Films

NIL is a promising technology that can fabricate high-resolution nanostructures beyond the limitation set by light diffraction in photolithography or beam scattering in charged particle beam lithography [15]. It has the capability to fabricate high-resolution nanostructures on large area of rigid or flexible substrates. Its unique advantages of low cost and high throughput are particularly desired by many emerging applications in the fields of flexible electronics, [23] photonics, [24] microfluidics, [25] photovoltaics, [26] just to name a few. In these emerging applications, plastic materials are widely used as device substrates mainly for their cost-effectiveness, flexibility, transparency, and lightweight, among other intriguing properties [27]. The desirable properties in plastic films to be used as device substrates are moderate mechanical strength, low surface energy, wide range optical transparency, chemical stability, processing convenience, and flexibility particularly when used in continuous roll-to-roll fabrication process [28].

Various polymeric materials have been used as substrates in flexible electronic devices, including but not limited to polyethylene terephthalate (PET), [29] polyethylenenapthalate (PEN), [30] polycarbonate (PC), [31] polyimide (PI), [32] polyetherimide (PEI), [33] polyethersulphone (PES), [34] poly(ether ketone) (PEK), [35] poly-tetrafluoro-ethylene (PTFE), [36] high density polyethylene (HDPE), [37] ethylene-tetrafluoroethylene (ETFE), [38] polyurethane acrylate (PUA), [39] polyvinyl alcohol (PVA), [40] and polyvinyl chloride (PVC) [41] etc. However, few of these materials possess a satisfactory combination of all the desirable properties, and there is still a demand for an ideal polymeric material that can be used as a superior alternative substrate material for flexible electronic devices.

COC is a relatively new family of plastic materials but increasingly getting popular [42]. Its unique combination of a number of promising properties such as low water absorption, good optical transparency in near-UV range, low surface energy, high strength, and high chemical resistance to acids and alkalis has made it an ideal material for optical devices [43] and microfluidic devices [44]. COC also possesses all the required properties for electronic devices. In particular, its flexibility makes it suitable to be used in highly flexible electronic devices, while its relatively high modulus and strong mechanical strength enable nanoscale features to be transferred with high fidelity.

A variety of fabrication approaches have been demonstrated on COC materials, including direct structuring methods (e.g., laser ablation [45] and micro-milling [46]) and replication techniques (e.g., hot embossing, [47] injection molding [48] and NIL [49]). Most of the reported replication methods so far are demonstrated with the fabrication of rather large structures (in range of micrometers) and many of these studies have mainly focused on device applications of the COC nanostructures [49, 50] or on optimizing the topography and surface properties of the master mold to enhance the morphology of the replicated COC nanostructures [51–53]. Systematic experimental study on the effects of thermal nanoimprint processing parameters on fabricating nanoscale features on COC films need to be studied.

To do so, we investigated the thermal nanoimprint process for replicating nanoscale structures into COC films. The effects of nanoimprint processing parameters on the mold cavity filling and pattern transfer fidelity are experimentally studied using grating structures of various sizes down to 70 nm half-pitch. The height and morphology of replicated gratings were analyzed by AFM and SEM characterization, and optimized processing parameters were determined for high pattern transfer fidelity and short nanoimprint processing time.

COC films (Grade 8007, Grade 6017) used in fabrication of EMTEs were obtained from TOPAS Advanced Polymers. These COC films were fabricated by extrusion process and their major properties are summarized in Table 3.1 (Appendix). The contact angle of distilled water on the COC film (shown in Fig. 2.10a) is measured to be 93° while optical transparency of blank COC film in UV and visible wavelength range (shown in Fig. 2.10b) is around 90% confirming the hydrophobic nature and superior optical transmission of COC films over the whole near-UV to near-infrared wavelength range.

Gratings of different periods of 420, 280, and 140 nm are replicated from silicon master molds to COC films by thermal NIL process. A home-built thermal nanoimprint setup was used to pattern the COC films with nanostructures. This setup consists

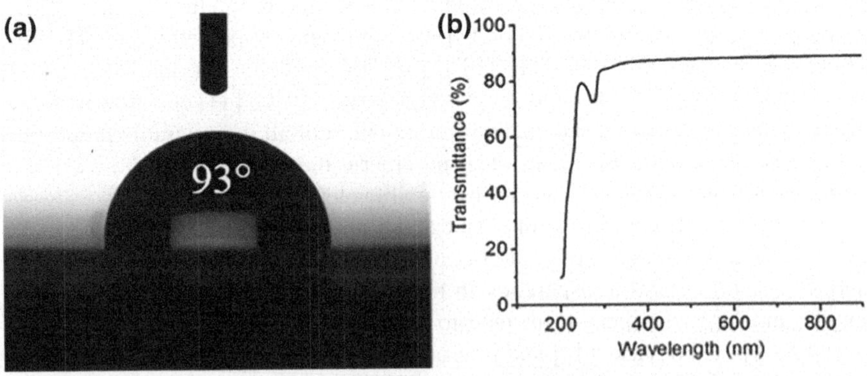

Fig. 2.10 Properties of the COC film **a** water contact angle measurement and **b** optical transparency in UV and visible wavelength range

of a manual hydraulic press, a set of electrically heated platens with temperature controller and a water circulation chiller. During the process, a silicon mold with a grating pattern was manually positioned onto the COC film placed at the center of the lower plate of the hydraulic press. Then the plates were heated to the required temperature and certain imprinting pressure was applied and held for 5 min. The heated platens were then cooled down to the demolding temperature of 50 °C and the COC film was peeled off from the silicon mold. The silicon molds used in our experiment have grating patterns with periods of 420 nm, 280 nm and 140 nm, and depths of 125 nm, 120 nm, and 70 nm, respectively. All gratings have a 50% duty cycle. In all the experiments, silicon molds were not treated with any anti-sticking layer, since the low surface energy of imprinted COC films ensured a reliable mold separation after processing.

The SEM and AFM images of master silicon grating molds with periods of 420, 280, and 140 nm are presented in Figs. 2.11a, 2.12a and 2.13a respectively. Similarly the SEM images and AFM images of successfully replicated nanogratings with

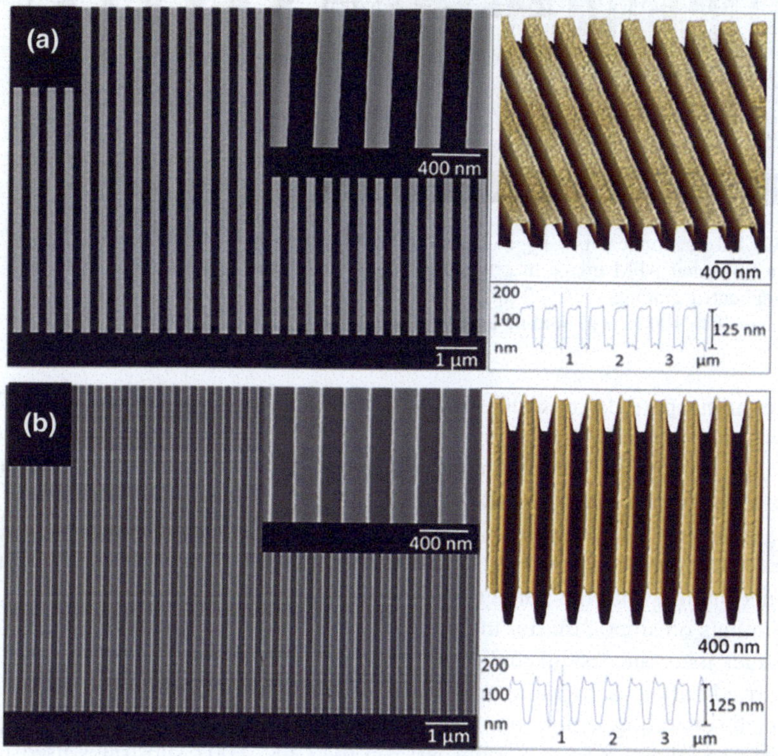

Fig. 2.11 Fabrication results using nanogratings of 420 nm period, 210 nm line width and 125 nm depth. **a** SEM and AFM image of original gratings on a silicon mold. **b** SEM and AFM image of the replicated gratings on COC (8007) film formed at imprint pressure P = 6.2 MPa, imprint temperature T = 100 °C and holding time t = 5 min

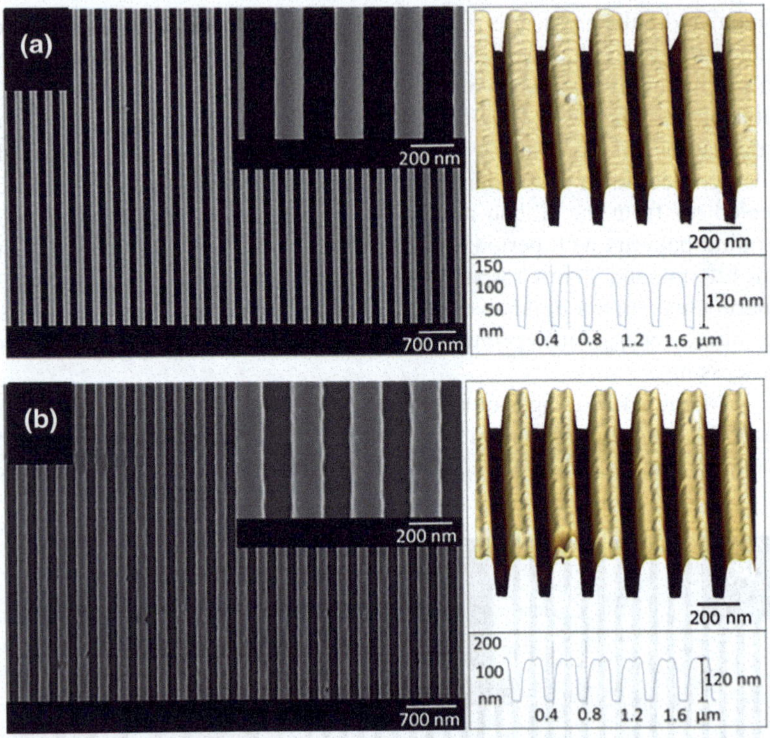

Fig. 2.12 Fabrication results using nanogratings of 280 nm period, 140 nm line width and 120 nm depth. **a** SEM and AFM image of original gratings on a silicon mold. **b** SEM and AFM image of the replicated gratings on COC (8007) film formed at imprint pressure P = 6.2 MPa, imprint temperature T = 100 °C and holding time t = 5 min

corresponding periods on COC films are shown in Figs. 2.11b, 2.12b and 2.13b respectively. Each SEM image presents the gratings at two different magnifications while the corresponding AFM image shows the 3D profile and the cross-sectional height profile of the gratings. All the imprinted results shown in these images were fabricated using optimal imprinting parameters (imprinting pressure P = 6.2 MPa, imprinting temperature T = 100 °C and holding time t = 5 min). It is clear from the presented images that the nanogratings in COC film are the complete copy of the nanogratings on silicon master mold.

Besides these successful replications, a series of nanoimprint experiments were also carried out on each grating mold using sub-optimized imprinting parameters to (1) find the operating region (minimum pressure and minimum temperature required for perfect replication) for thermal NIL on the COC 8007 substrates used in this work, and (2) investigate the imprinted height as a function of imprinting parameters during the thermal NIL process. For each set of experiments, the imprinted height was measured by AFM and compared to the depth of the trench on the silicon mold.

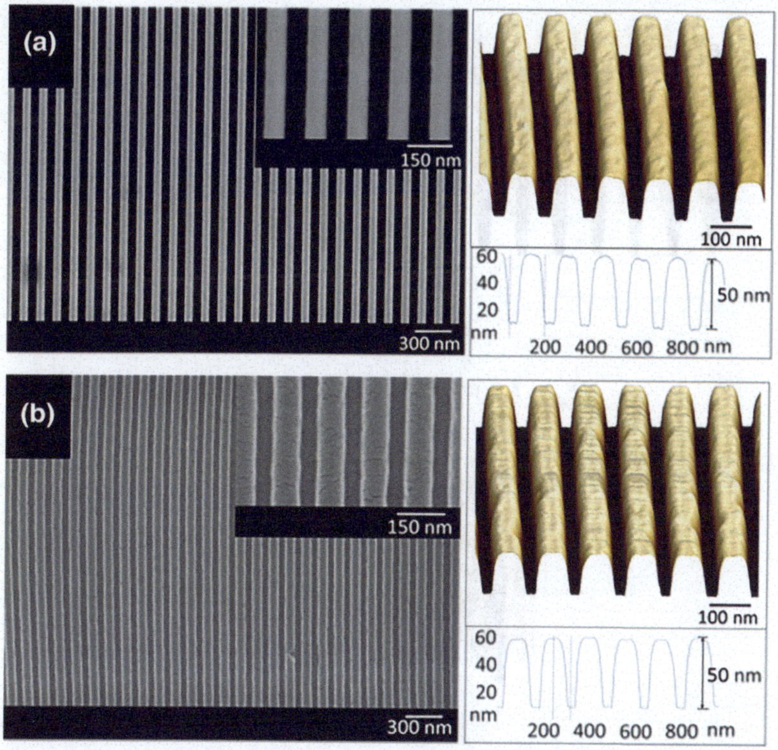

Fig. 2.13 Fabrication results using nanogratings of 140 nm period, 70 nm line width and 50 nm depth. **a** SEM and AFM image of original gratings on a silicon mold. **b** SEM and AFM image of the replicated gratings on COC (8007) film formed at imprint pressure P = 6.2 MPa, imprint temperature T = 100 °C and holding time t = 5 min

Typical AFM images of partially filled gratings with periods of 420, 280, 140 nm are shown in Fig. 2.14a, b, and c respectively.

The summary of imprinting parameters for incomplete and complete replications of all three periods of nanogratings is shown in Fig. 2.15a. It can be concluded from Fig. 2.15a that irrespective of grating size, the minimum imprinting pressure of 3 MPa and imprinting temperature of 85 °C is required for perfect replication through thermal NIL on COC 8007 substrates. However, it is clear from the quantitative data presented in Fig. 2.15b, c and d that, under same sub-optimal imprinting parameters, different height is replicated for each grating mold. For instance, at imprinting pressure of 12.55 MPa and imprinting temperature of 80 °C, the replicated heights for 140 nm period gratings, 280 nm period gratings and 420 nm period are 30 nm, 70 nm and 88 nm respectively. These incompletely replicated results indicate that the grating replication process is more influenced by imprinting temperature as compared to imprinting pressure. The reason for the phenomenon is that the modulus of the COC film abruptly changes near its glass transition temperature due to its viscoelastic behavior. At a temperature well below glass transition temperature, COC is in glassy

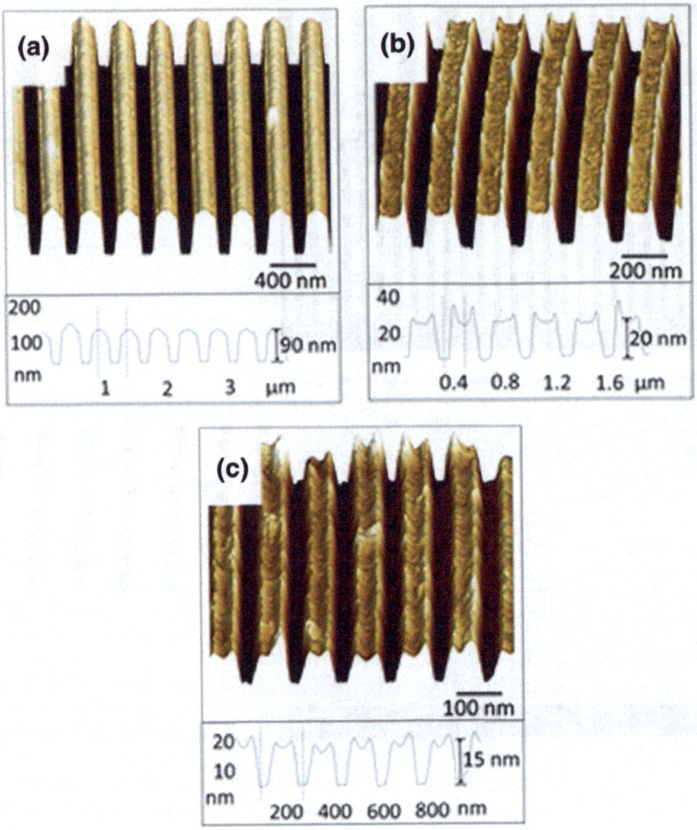

Fig. 2.14 AFM images of typical COC gratings imprinted using sub-optimal processing parameters. **a** Gratings having 420 nm period, 210 nm line width and 90 nm depth; **b** gratings having 280 nm period, 140 nm line width and 20 nm depth; and **c** gratings having 140 nm period, 70 nm line width and 15 nm depth

region where its Young's modulus is high and remains constant over a wide range of temperature. The glassy COC film appears rigid and deforms in a more elastic manner. However, with increasing temperature over glass transition temperature, the thermal vibration becomes intense enough to overcome the potential barriers for rotation and translation of COC molecule segments [54]. Therefore, the modulus of COC at the transition region decreases by several orders of magnitude, [55] leading to a more pronounced effect of imprinting temperature over that of imprinting pressure.

When used as substrate for flexible electronic devices, the COC is expected to have a high modulus therefore minimum degradation during the bending. The modulus of COC film with nanostructures is experimentally investigated. Figure 2.16 shows the AFM based measurement on the elastic modulus distribution of imprinted 420 nm period COC gratings together with the corresponding morphology image.

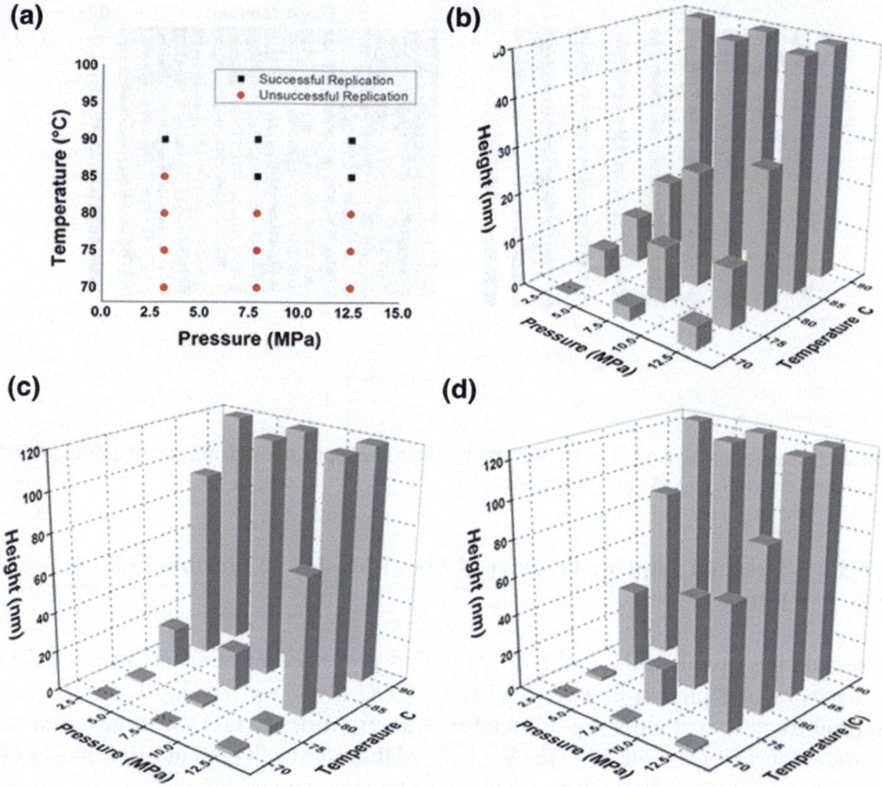

Fig. 2.15 Influence of processing parameters on the pattern fidelity. **a** Summary of results for incomplete and complete replication of the gratings (irrespective of grating period). **b–d** Quantitative relationship between the depth of imprinted patterns, imprinting temperature, and imprinting pressure for 140 nm (50 nm height), 280 nm (120 nm height) and 420 nm (125 nm height) gratings, respectively. All grating molds have a 1:1 duty cycle

The modulus mapping was obtained through the PeakForce Quantitative Nanomechanical Property Module on the AFM using a Derjaguin Muller Toporov (DMT) model [56]. The DMT elastic modulus measured at the center of the grating lines and at the center of the grating trenches ranges from 5 to 9 GPa, which is very high compared with other plastic materials used previously for nanoimprint templates. At the edge of grating lines, the AFM measurement cannot provide a reliable modulus value due to the effect of grating sidewalls. It is worth to note that, the DMT modulus measured on the grating lines and in the trenches is apparently higher than the modulus measured on a featureless COC film, which is ~2.7 GPa. This increase in the modulus presumably comes from the thermal NIL process where the film experiences a high compressive stress which in turn increases the molecular density and hence the modulus of the film, [57] but further study on this phenomenon and its effect on nanoimprint applications is needed.

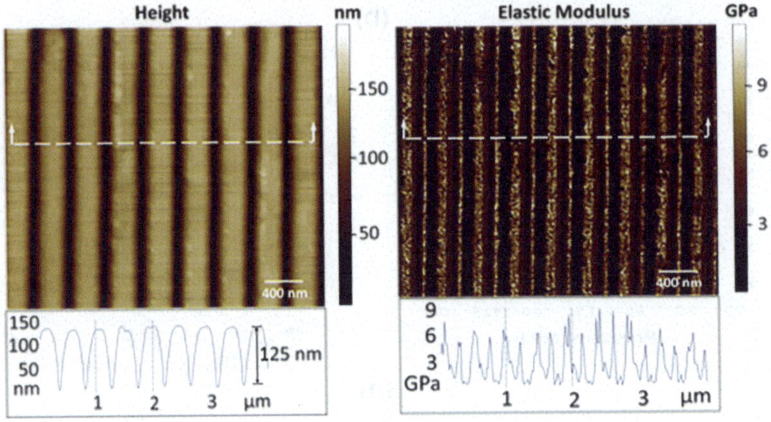

Fig. 2.16 Representative topography and DMT modulus mappings of 420 nm period gratings COC film using PeakForce QNM imaging mode

2.3.3.2 Thermal Imprint Transfer of Metal-Meshes to Flexible COC Films

Based on the excellent thermal NIL results (as presented in previous Sect. 2.3.3.1) in terms of nanopattern transfer fidelity on COC films, we developed the thermal imprint transfer methodology of metal-meshes from rigid glass substrates to these films, to fabricate flexible EMTEs. We utilized this approach both in LEIT and TEIT fabrication strategies for making micro-EMTEs and nano-EMTEs.

In both LEIT and TEIT techniques, the metal-meshes were transferred to 100-µm-thick COC films (Grade 8007, Grade 6017) using a home-built setup consisting of a hydraulic press, electrically heated platens with a temperature controller, and a chiller. During the thermal imprint transfer process, the plates were heated to the required temperature and an imprinting pressure was applied. The heated platens were then cooled to the demolding temperature. Finally, the COC film was peeled from the glass substrate, with the metal-mesh fully embedded in the COC film. For instance, Fig. 2.17 describes the metal-mesh transfer to COC film using thermal imprint transfer in TEIT fabrication strategy. Further details of the metal-mesh transfers using thermal imprint process are documented in Chaps. 3, 4 and 5 of this dissertation.

2.3.3.3 Motivation from UV-NIL on Epoxy Coated COC Films for UV Imprint Transfer of Metal-Meshes to Epoxy Coated COC Films

The successfully fabricated nanogratings on COC film as explained in previous section were used as a template to replicate nanogratings into UV curable epoxy by UV-NIL process. In this investigation, we used UV-curable epoxy (NOA-61) as

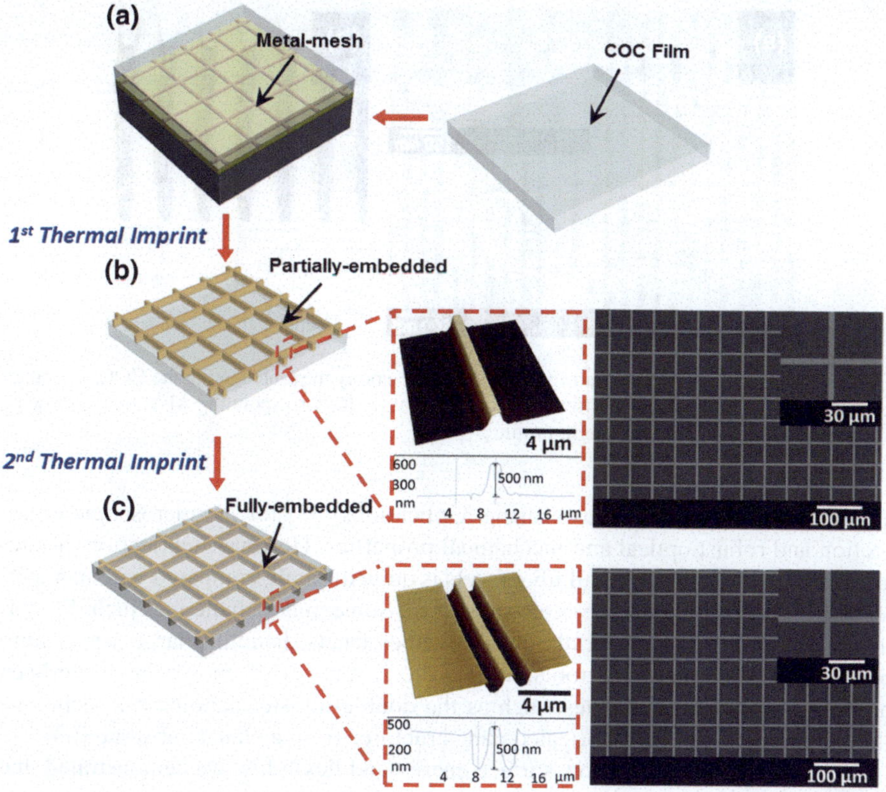

Fig. 2.17 Schematic illustrations and morphological characterizations (AFM (left) and SEM (right)) of the thermal imprint transfer in TEIT fabrication process **a** heating and pressing the metal mesh into a COC film, **b** peeling off the COC film with the metal-mesh transferred in a partially embedded form **c** second heating and pressing the metal mesh into the COC film

the UV nanoimprint resist. The SEM and AFM images of 140 nm period gratings transferred onto the epoxy layer are shown in Fig. 2.18a and b, respectively. COC templates were tested for up to ten imprinting cycles and by investigating both the replicated grating profiles on the epoxy and the surface of the used template, we found no noticeable defects, surface deterioration, contamination, swelling, deformation and collapse of features. This indicates that the COC templates are reasonably robust and durable. The well-replicated grating profiles on the epoxy for a large number of imprinting cycles also demonstrate the feasibility of nanostructured COC films as low-cost secondary NIL templates for transferring sub-100 nm features. Moreover, these flexible templates can be utilized in roll–to-roll UV NIL process by wrapping it around an imprinting roller.

We also demonstrated transferring metallic nanostructures onto very small size substrates i.e. optical fiber facets using COC secondary templates. Optical fiber is

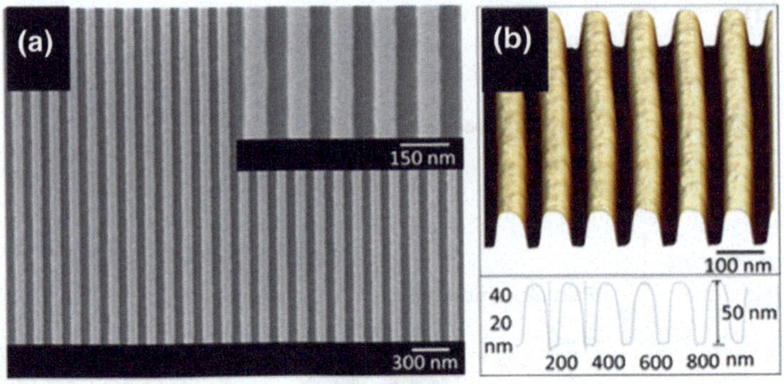

Fig. 2.18 140 nm period nanogratings on UV curable epoxy imprinted using the COC secondary template. **a** SEM images at two different magnifications, **b** the corresponding AFM image showing 3D profile and cross-sectional height profile

a unique platform for various sensing applications [58] for its microscopic cross-section and robust optical and mechanical properties. However, fabrication of integrated nanostructures on small fiber facets is quite difficult, compared with nanopatterning on conventional large-scale substrates. Some major challenges include coating of thin-film resist materials on small fiber facets, handling large-aspect-ratio optical fibers in processing tools, and so on. A number of approaches have been proposed to solve these issues, such as the double-transfer nanoimprint technique [59] and transferring pre-fabricated nanostructures from a planar substrate [60].

Because of its low cost, low surface energy and flexibility, we demonstrated that COC templates can be an ideal choice for transferring metallic nanostructures onto optical fiber facets. It's advantageous to use COC as the underlying substrate for the metal structure because the adhesion between them is typically weak. On the other hand, since the epoxy is also in partial contact with the underlying substrate during the metal transfer process, the weak adhesion between the COC substrate and the epoxy ensures an easy separation between the cured epoxy and the COC substrate. Figure 2.19a and b show the SEM and AFM images of successfully fabricated 420 nm period gold grating on 125 μm diameter optical fiber facet. The gold grating has a line width of 210 nm and a height of 125 nm. These metallic nanostructures can be exploited to allow strong optical coupling of incident light to localized surface plasmons. With further design and optimization of the metallic structures to largely enhance local electrical field, such metallic nanostructures integrated with optical fibers can function as high-performance sensors for surface-enhanced spectroscopy techniques [61].

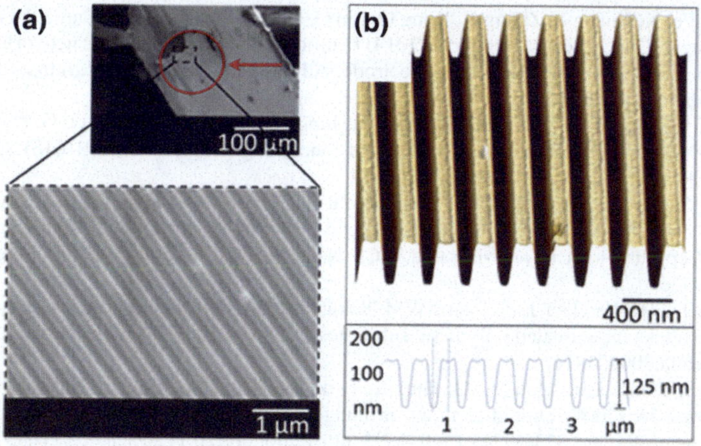

Fig. 2.19 420 nm period gold grating structures transferred onto fiber facet. **a** SEM images at different magnifications showing the optical fiber tip and the transferred gratings. Red arrow indicates the position of the optical fiber facet. **b** The corresponding AFM image showing 3D profile and cross-sectional height profile of the gratings

2.3.3.4 UV-Imprint Transfer of Metal-Meshes to Epoxy Coated COC Films

Based on the promising UV-NIL results (as presented in previous Sect. 2.3.3.3) in terms of nanopattern transfer fidelity on epoxy/COC substrates, we established the UV-imprint transfer methodology for metal-mesh transfers from electrodeposition templates to these films. We utilized this process in fabrication of nano-EMTEs by transferring nano-meshes onto UV-curable epoxy coated COC films. In practice, a UV-curable epoxy was drop-casted on the electroplated silica-template and then COC substrate was placed on top of it. The UV-curable epoxy/COC substrate was cured by a UV light along with a small pressure to press the template and the substrate. Finally after curing, the epoxy/COC substrate was manually peeled off from the silica-template, transferring along the metal nano-mesh in embedded form. More details of the UV-imprint process for metal-mesh transfer are presented in Chap. 5 of this thesis.

References

1. Wu H, Kong D, Ruan Z, Hsu P-C, Wang S, Yu Z, Carney TJ, Hu L, Fan S, Cui Y (2013) A transparent electrode based on a metal nanotrough network. Nat Nano 8(6):421–425. https://doi.org/10.1038/nnano.2013.84http://www.nature.com/nnano/journal/v8/n6/abs/nnano.2013.84.html#supplementary-information

2. Han B, Pei K, Huang Y, Zhang X, Rong Q, Lin Q, Guo Y, Sun T, Guo C, Carnahan D, Giersig M, Wang Y, Gao J, Ren Z, Kempa K (2014) Uniform self-forming metallic network as a high-performance transparent conductive electrode. Adv Mater 26(6):873–877. https://doi.org/10.1002/adma.201302950
3. Kim H-J, Lee S-H, Lee J, Lee E-S, Choi J-H, Jung J-H, Jung J-Y, Choi*, D.-G. (2014) High-durable AgNi nanomesh film for a transparent conducting electrode. Small 10(18):3767–3774. https://doi.org/10.1002/smll.201400911
4. Bao C, Yang J, Gao H, Li F, Yao Y, Yang B, Fu G, Zhou X, Yu T, Qin Y, Liu J, Zou Z (2015) In Situ fabrication of highly conductive metal nanowire networks with high transmittance from deep-ultraviolet to near-infrared. ACS Nano 9(3):2502–2509. https://doi.org/10.1021/nn504932e
5. van Osch THJ, Perelaer J, de Laat AWM, Schubert US (2008) Inkjet printing of narrow conductive tracks on untreated polymeric substrates. Adv Mat 20(2):343–345. https://doi.org/10.1002/adma.200701876
6. Ahn BY, Duoss EB, Motala MJ, Guo X, Park S-I, Xiong Y, Yoon J, Nuzzo RG, Rogers JA, Lewis JA (2009) Omnidirectional printing of flexible, stretchable, and spanning silver microelectrodes. Sci 323(5921):1590–1593. https://doi.org/10.1126/science.1168375
7. Khan A, Rahman K, Hyun M-T, Kim D-S, Choi K-H (2011) Multi-nozzle electrohydrodynamic inkjet printing of silver colloidal solution for the fabrication of electrically functional microstructures. Appl Phys A 104(4):1113–1113. https://doi.org/10.1007/s00339-011-6386-0
8. Khan A, Rahman K, Kim DS, Choi KH (2012) Direct printing of copper conductive micro-tracks by multi-nozzle electrohydrodynamic inkjet printing process. J Mater Proc Technol 212(3):700–706. https://doi.org/10.1016/j.jmatprotec.2011.10.024
9. Ellmer K (2012) Past achievements and future challenges in the development of optically transparent electrodes. Nat Photon 6(12):809–817
10. Choi H-J, Choo S, Jung P-H, Shin J-H, Kim Y-D, Lee H (2015) Uniformly embedded silver nanomesh as highly bendable transparent conducting electrode. Nanotechnol 26(5):055305
11. Hierlemann A, Brand O, Hagleitner C, Baltes H (2003) Microfabrication techniques for chemical/biosensors. Proceedings of the IEEE 91(6):839–863. https://doi.org/10.1109/JPROC.2003.813583
12. Berger SD, Gibson JM, Camarda RM, Farrow RC, Huggins HA, Kraus JS, Liddle JA (1991) Projection electron-beam lithography: a new approach. J Vac Sci Technol B 9(6):2996–2999. https://doi.org/10.1116/1.585356
13. Whang D, Jin S, Lieber CM (2003) Nanolithography using hierarchically assembled nanowire masks. Nano Lett 3(7):951–954. https://doi.org/10.1021/nl034268a
14. Colli A, Fasoli A, Pisana S, Fu Y, Beecher P, Milne WI, Ferrari AC (2008) Nanowire lithography on silicon. Nano Lett 8(5):1358–1362. https://doi.org/10.1021/nl080033t
15. Chou SY, Krauss PR, Renstrom PJ (1995) Imprint of sub-25 nm vias and trenches in polymers. Appl Phys Lett 67(21):3114–3116. https://doi.org/10.1063/1.114851
16. Moon Kyu K, Jong GO, Jae Yong L, Guo LJ (2012) Continuous phase-shift lithography with a roll-type mask and application to transparent conductor fabrication. Nanotechnology 23(34):344008
17. Manfrinato VR, Zhang L, Su D, Duan H, Hobbs RG, Stach EA, Berggren KK (2013) Resolution limits of electron-beam lithography toward the atomic scale. Nano Lett 13(4):1555–1558. https://doi.org/10.1021/nl304715p
18. Yu F, Li P, Shen H, Mathur S, Lehr C-M, Bakowsky U, Mücklich F (2005) Laser interference lithography as a new and efficient technique for micropatterning of biopolymer surface. Biomaterials 26(15):2307–2312. https://doi.org/10.1016/j.biomaterials.2004.07.021
19. Garcia R, Knoll AW, Riedo E (2014) Advanced scanning probe lithography. Nat Nano 9(8):577–587. https://doi.org/10.1038/nnano.2014.157
20. Schwarzacher W (2006) Electrodeposition: A technology for the future. Electrochem Soc Interface 1:4
21. Djokić SS, Cavallotti PL (2010) Electroless Deposition: Theory and Applications. In: Djokic SS (eds) In electrodeposition: theory and practice. Springer, New York, pp 251–289

22. Khan A, Li S, Tang X, Li W-D (2014) Nanostructure transfer using cyclic olefin copolymer templates fabricated by thermal nanoimprint lithography. J Vac Sci Technol B 32(6):06FI02. https://doi.org/10.1116/1.4900609
23. McAlpine MC, Friedman RS, Lieber CM (2003) Nanoimprint lithography for hybrid plastic electronics. Nano Lett 3:443
24. Li WD, Ding F, Hu J, Chou SY (2011) Three-dimensional cavity nanoantenna coupled plasmonic nanodots for ultrahigh and uniform surface-enhanced Raman scattering over large area. Opt Express 19:3925. https://doi.org/10.1364/OE.19.003925
25. Guo LJ, Cheng X, Chou CF (2003) Fabrication of size-controllable nanofluidic channels by nanoimprinting and its application for DNA stretching. Nano Lett 4:69. https://doi.org/10.1021/nl034877i
26. Cheyns D, Vasseur K, Rolin C, Genoe J, Poortmans J, Heremans P (2008) Nanoimprinted semiconducting polymer films with 50 nm features and their application to organic heterojunction solar cells. Nanotechnol 19:424016
27. Ahn SH, Guo LJ (2008) High-Speed roll-to-roll nanoimprint lithography on flexible plastic substrates. Adv Mater 20:2044. https://doi.org/10.1002/adma.200702650
28. Søndergaard R, Hösel M, Angmo D, Larsen-Olsen TT, Krebs FC (2012) Roll-to-roll fabrication of polymer solar cells. Mat Today 15(1):36–49. https://doi.org/10.1016/S1369-7021(12)70019-6
29. Yu S, Han HJ, Kim JM, Yim S, Sim DM, Lim H, Lee JH, Park WI, Park JH, Kim KH, Jung YS (2017) Area-selective lift-off mechanism based on dual-triggered interfacial adhesion switching: highly facile fabrication of flexible nanomesh electrode. ACS Nano 11(4):3506–3516. https://doi.org/10.1021/acsnano.7b00229
30. Fonrodona M, Escarré J, Villar F, Soler D, Asensi JM, Bertomeu J, Andreu J (2005) PEN as substrate for new solar cell technologies. Sol Energy Mater Sol Cells 89(1):37–47. https://doi.org/10.1016/j.solmat.2004.12.006
31. Fateh R, Ismail AA, Dillert R, Bahnemann DW (2011) Highly active crystalline mesoporous TiO2 films coated onto polycarbonate substrates for self-cleaning applications. J Phys Chem C 115(21):10405–10411. https://doi.org/10.1021/jp200892z
32. Gao X, Lin L, Liu Y, Huang X (2015) LTPS TFT process on polyimide substrate for flexible AMOLED. J Disp Technol 11(8):666–669. https://doi.org/10.1109/JDT.2015.2419656
33. Ok K-H, Kim J, Park S-R, Kim Y, Lee C-J, Hong S-J, Kwak M-G, Kim N, Han CJ, Kim J-W (2015) Ultra-thin and smooth transparent electrode for flexible and leakage-free organic light-emitting diodes. 5:9464. https://doi.org/10.1038/srep09464, https://www.nature.com/articles/srep09464#supplementary-information
34. Qin F, Tong J, Ge R, Luo B, Jiang F, Liu T, Jiang Y, Xu Z, Mao L, Meng W, Xiong S, Li Z, Li L, Zhou Y (2016) Indium tin oxide (ITO)-free, top-illuminated, flexible perovskite solar cells. J Mater Chem A 4(36):14017–14024. https://doi.org/10.1039/C6TA06657G
35. Wang Y, Chen B, Evans KE, Ghita O (2016) Novel fibre-like crystals in thin films of Poly Ether Ether Ketone (PEEK). Mater Lett 184:112–118. https://doi.org/10.1016/j.matlet.2016.08.024
36. Wittmann JC, Smith P (1991) Highly oriented thin films of poly(tetrafluoroethylene) as a substrate for oriented growth of materials. Nature 352(6334):414–417
37. Zheng Q, Peng M, Yi X (1999) Crystallization of high density polyethylene: effect of contact with NdFeB magnetic powder substrates. Mat Lett 40(2):91–95. https://doi.org/10.1016/S0167-577X(99)00054-3
38. Ahn SH, Guo LJ (2009) Large-area roll-to-roll and roll-to-plate nanoimprint lithography: a step toward high-throughput application of continuous nanoimprinting. ACS Nano 3:2304
39. Seo SM, Kim TI, Lee HH (2007) Simple fabrication of nanostructure by continuous rigiflex imprinting. Microelectron Eng 84:567. https://doi.org/10.1016/j.mee.2006.11.008
40. Schaper CD (2003) Patterned transfer of metallic thin film nanostructures by water-soluble polymer templates. Nano Lett 3:1305. https://doi.org/10.1021/nl034412s
41. Hong SH, Hwang JY, Lee H, Lee HC, Choi KW (2009) UV nanoimprint using flexible polymer template and substrate. Microelectron Eng 86:295

42. Lee H, Hong S, Yang K, Choi K (2006) Fabrication of 100nm metal lines on flexible plastic substrate using ultraviolet curing nanoimprint lithography. Appl Phys Lett 88:143112. https://doi.org/10.1063/1.2193653
43. Okagbare PI, Emory JM, Datta P, Goettert J, Soper SA (2010) Fabrication of a cyclic olefin copolymer planar waveguide embedded in a multi-channel poly (methyl methacrylate) fluidic chip for evanescence excitation. Lab Chip 10:66
44. Nunes PS, Ohlsson PD, Ordeig O, Kutter JP (2010) Cyclic olefin polymers: emerging materials for lab-on-a-chip applications. Microfluid Nanofluidics 9:145
45. Bundgaard F, Perozziello G, Geschke O (2006) Rapid prototyping tools and methods for all-Topas® cyclic olefin copolymer fluidic microsystems. P I Mech Eng C-J Mech 220:1625
46. Steigert J, Haeberle S, Brenner T, Müller C, Steinert C, Koltay P, Gottschlich N, Reinecke H, Rühe J, Zengerle R (2007) Rapid prototyping of microfluidic chips in COC. J Micromech Microeng 17:333
47. Leech P (2009) Hot embossing of cyclic olefin copolymers. J Micromech Microeng 19:055008
48. Kalima V, Pietarinen J, Siitonen S, Immonen J, Suvanto M, Kuittinen M, Mönkkönen K, Pakkanen T (2007) Transparent thermoplastics: Replication of diffractive optical elements using micro-injection molding. Opt Mater 30:285
49. Malic L, Cui B, Tabrizian M, Veres T (2009) Nanoimprinted plastic substrates for enhanced surface plasmon resonance imaging detection. Opt Express 17:20386
50. Vannahme C, Klinkhammer S, Christiansen MB, Kolew A, Kristensen A, Lemmer U, Mappes T (2010) All-polymer organic semiconductor laser chips: parallel fabrication and encapsulation. Opt Express 18:24881
51. Matschuk M, Larsen NB (2013) Injection molding of high aspect ratio sub-100 nm nanostructures. J Micromech Microeng 23:025003
52. Pakkanen TT, Hietala J, Pääkkönen EJ, Pääkkönen P, Jääskeläinen T, Kaikuranta T (2002) Replication of sub-micron features using amorphous thermoplastics. Polym Eng Sci 42:1600
53. Gadegaard N, Mosler S, Larsen NB (2003) Biomimetic polymer nanostructures by injection molding. Macromol Mater Eng 288:76
54. Shaw MT, MacKnight WJ (2005) Introduction to polymer viscoelasticity. John Wiley & Sons, New Jersey
55. Jena R, Chen X, Yue C, Lam Y (2011) Rheological (visco-elastic behaviour) analysis of cyclic olefin copolymers with application to hot embossing for microfabrication. J Micromech Microeng 21:085029
56. Derjaguin BV, Muller VM, Toporov YP (1975) Effect of contact deformations on the adhesion of particles. J Colloid Interf Sci 53:314. https://doi.org/10.1016/0021-9797(75)90018-1
57. Woo YS, Kim JK, Lee DE, Suh KY, Lee WI (2007) Density variation of nanoscale patterns in thermal nanoimprint lithography. Appl Phys Lett 91:253111. https://doi.org/10.1063/1.2827187
58. Kostovski G, Stoddart PR, Mitchell A (2014) The optical fiber tip: an inherently light-coupled microscopic platform for micro-and nanotechnologies. Adv Mater 26:3798
59. Shen Y, Yao L, Li Z, Kou J, Cui Y, Bian J, Yuan C, Ge H, Li WD, Wu W, Chen Y (2013) Double transfer UV-curing nanoimprint lithography. Nanotechnology 24:465304
60. Scheerlinck S, Dubruel P, Bienstman P, Schacht E, Van Thourhout D, Baets R (2009) Metal grating patterning on fiber facets by UV-based nano imprint and transfer lithography using optical alignment. J Lightwave Technol 27:1415
61. Yang X, Ileri N, Larson CC, Carlson TC, Britten JA, Chang ASP, Gu C, Bond TC (2012) Nanopillar array on a fiber facet for highly sensitive surface-enhanced Raman scattering. Opt Express 20:24819. https://doi.org/10.1364/OE.20.024819

Chapter 3
Micro Embedded Metal-mesh Transparent Electrodes (Micro-EMTEs) Fabricated by LEIT Strategy

Abstract This chapter presents the LEIT fabrication for micro-EMTEs in detail. The LEIT fabrication approach replaces vacuum-based metal deposition with an electrodeposition and is potentially suitable for high-throughput, large-volume and low-cost production. This approach can be easily adapted to make flexible and even stretchable devices Jang et al. (IEEE transaction on Antenna and Propagation, 2015 [1]). Prototype copper micro-EMTEs are fabricated on flexible COC films with superior electrical conductivity and optical transmittance. FoM (σ_{dc}/σ_{opt}) values as high as 1.5×10^4 have been demonstrated on the sample copper micro-EMTEs. This fabrication process has been demonstrated to be able to scale for a larger EMTE area and a wide range of materials. Because of the electrode's embedded nature, excellent mechanical, chemical, and environmental stability were observed

3.1 Introduction to LEIT Fabrication Strategy

This chapter presents one of our cost-effective and solution processed fabrication strategy for EMTEs; combining lithography, electroplating, and imprint transfer. This LEIT fabrication approach replaces vacuum-based metal deposition with an electrodeposition and is potentially suitable for high-throughput, large-volume and low-cost production. In particular, this strategy enables fabrication of a high-aspect-ratio (thickness to linewidth) metal mesh, substantially improving conductivity without considerably sacrificing transparency.

The fabrication process is schematically described in Fig. 3.1. In a typical fabrication process, after a photoresist layer is spincoated on a cleaned conductive glass substrate (Fig. 3.1a), a photolithography step is conducted to create a mesh pattern in the photoresist through ultraviolet exposure and development (Fig. 3.1b), exposing the conductive glass surface in the mesh trench. In the following electrodeposition step (Fig. 3.1c), selected metal is deposited inside the lithographically defined trenches and fills the trenches with a uniform metal mesh. Next, the photoresist is gently dissolved in solvent, leaving the bare metal mesh on the surface of the conductive glass (Fig. 3.1d). A thermoplastic film is then placed on the metal mesh and heated to above its glass transition temperature. A uniform pressure is applied

A. Khan, *Novel Embedded Metal-mesh Transparent Electrodes*, Springer Theses,
https://doi.org/10.1007/978-981-15-2918-4_3

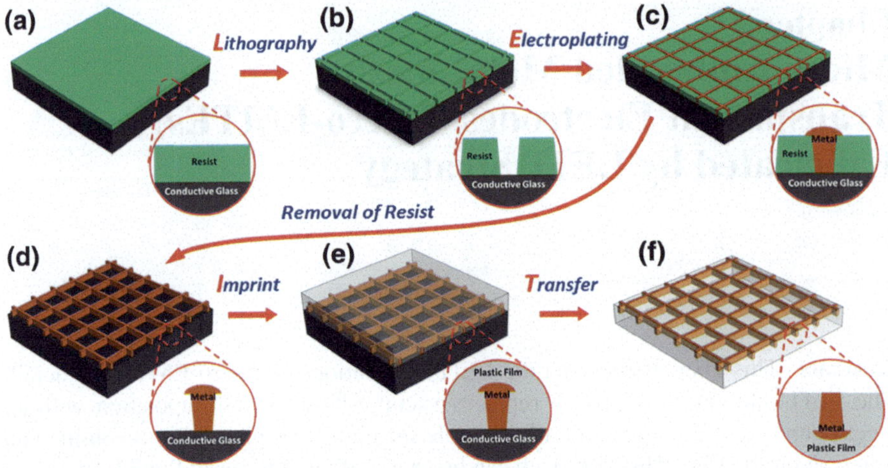

Fig. 3.1 Schematic illustrations of LEIT fabrication process. **a** Photoresist layer on conductive glass substrate formed by spin coating. **b** Mesh patterns formed in a photoresist layer coated by photolithography. **c** Electrodeposition of metal inside the resist trenches to form a uniform metal mesh. **d** Removal of photoresist to obtain standing bare metal mesh on the conductive glass substrate. **e** Heating and pressing the metal mesh into a plastic COC film. **f** Peeling off the plastic COC film with the metal mesh transferred in a fully embedded form

to imprint the metal mesh into the softened plastic film (Fig. 3.1e). Finally, after cooling the stack and separating the plastic film from the conductive glass, the metal mesh is transferred and embedded in the plastic film to complete the fabrication process (Fig. 3.1f). The entire fabrication process is solution-based and performed in an ambient environment without any vacuum processing, and it can thus be readily adapted to large-volume production.

3.2 Experimental Details

3.2.1 Fabrication of Flexible Micro-EMTEs by LEIT Strategy

First, FTO glass substrates (approximately 15 Ω/sq, South China Xiang S&T, China) were cleaned with a cotton swab and liquid detergent, rinsed thoroughly with another cotton swab and DI water, and then further cleaned by ultrasonication in isopropanol and DI water for 30 s before being dried with compressed air. AZ 1500 (Clariant, Switzerland) photoresist was spin-coated at 4000 rpm for 60 s to reach a film thickness of 1.8 μm on the cleaned FTO glass. It was then baked on a hotplate at 100 °C for 50 s. Thereafter, the photoresist was exposed using a URE-2000/35 UV mask aligner (Chinese Academy of Sciences, China) for an exposure dose of 20 mJ/cm^2. The photoresist was then developed in AZ 300 MIF developer (Clariant, Switzerland) for 50 s. The samples were finally rinsed in DI water and blow-dried with compressed air.

A subsequent electrodeposition process used commercial aqueous copper, silver, gold, nickel, and zinc plating solutions (Caswell, USA). A Keithley 2400 SourceMeter was used to supply a constant 5 mA current to a two-electrode electrodeposition setup with the photoresist-covered FTO glass as the working electrode and metal bar as the counter electrode. Electrodeposition time was varied to achieve the desired thickness of deposited metal. After the electrodeposition was completed, the sample was thoroughly rinsed with DI water, blow-dried with compressed air, and then placed in acetone for 5 min to remove the photoresist, leaving the bare metal mesh on the FTO glass substrate.

The metal mesh was then transferred to a 100 μm-thick COC film (Grade 8007) by thermal imprint using a home-built setup consisting of a hydraulic press (Specac Ltd., UK), electrically heated platens with a temperature controller (Specac Ltd., UK), and a chiller (Grant Instruments, UK). During the thermal imprint process, the plates were heated to 100 °C and an imprinting pressure of 15 MPa was applied, holding it for 5 min. The heated platens were then cooled to the demolding temperature of 40 °C. Finally, the COC film was peeled from the FTO glass, with the metal mesh fully embedded in the COC film.

3.2.2 Performance Characterizations

The morphology of the samples was characterized using an S-3400 N SEM (Hitachi, Japan) and a Multimode-8 AFM (Bruker, USA). The sheet resistance of the EMTE samples was measured using the four-probe method to eliminate contact resistance. During the measurement, four probes were placed on two silver paste-covered edges of a square sample, and the resistance was recorded with a Keithley 2400 SourceMeter (Keithley, USA). Optical transmission spectra were recorded using a Lambda 25 UV–Vis spectrometer (Perkin Elmer, USA). All transmittance values presented in this paper are normalized to the absolute transmittance through the bare COC film substrate. Energy-dispersive X-ray spectroscopy (EDS) analysis was performed by S-3400 N scanning electron microscope (Hitachi, Japan).

3.3 Experimental Results

3.3.1 *Morphological Characterizations of the LEIT Fabricated Micro-EMTEs*

An EMTE with a copper mesh on a COC film was fabricated through LEIT process as a demonstration. Using copper in TCs is advantageous because of copper's low resistivity and high abundance [2–4]. COC is selected as the substrate material because of its unique combination of beneficial properties as discussed in chap. 2 (Sect. 2.3). In addition, COC exhibits particularly favorable optical transparency extending into the near-UV range (Fig. 3.10, Appendix A), which is desirable in photovoltaic applications. SEM and AFM images in Fig. 3.2a–c show the morphological characterization of the micro-EMTE at different stages of the fabrication. Figure 3.2a displays the SEM and AFM images of the trench in the photoresist film created using photolithography (Fig. 3.1b). On this sample, the photoresist trench had a 50 μm pitch, and its trench width and depth were approximately 4 and 2 μm, respectively. Figure 3.2b presents the electroplated copper mesh on the FTO glass (Fig. 3.1d). The copper mesh was deposited with a 5 mA current over a 2 × 2 cm area. As evident from the images, the copper mesh had linewidth and thickness of approximately 4 and 1.8 μm, respectively. Figure 3.2c reveals that the copper mesh finally transferred onto a COC film (Fig. 3.1f). The AFM characterization revealed that the surface roughness of the completed EMTE with a 1.8 μm mesh thickness was less than 50 nm, confirming its fully embedded structure.

The fabrication process was further investigated by changing the electrodeposition time to fabricate copper micro-EMTEs of varying thickness, in which the electrodeposition current (5 mA) and substrate size (2 × 2 cm) are maintained constant. The relationship of metal thickness and electrodeposition time is summarized in Fig. 3.3. The curve indicates that metal thickness increases nonlinearly with electrodeposition time. This is due to the cross-section of the trench not being perfectly rectangular (Fig. 3.2a); rather, it is narrower at the bottom. Therefore, during electrodeposition with a constant current, the rate of increase in deposited metal thickness (i.e., the slope of the curve in Fig. 3.3) decreases with time. Thus, the electrodeposited metal mesh has a larger width at the upper part, which is useful for imprint transfer because it can be mechanically anchored in the substrate. With a sufficiently long electrodeposition time, the overplating of metal out of the trench results in a slower increase rate of metal thickness and a further increased lateral dimension on the overplated cap, which was the case for the two samples with 15 and 18 min of electrodeposition time in Fig. 3.3. Detailed characterizations of these samples can be found in the Appendix (Fig. 3.11). After removal of the photoresist template, these metal meshes were transferred to COC films through thermal imprinting. However, the transfer could be successful only for meshes thicker than 600 nm. The reason for the unsuccessful transfer was that the COC film trapping force applied on the sidewall of

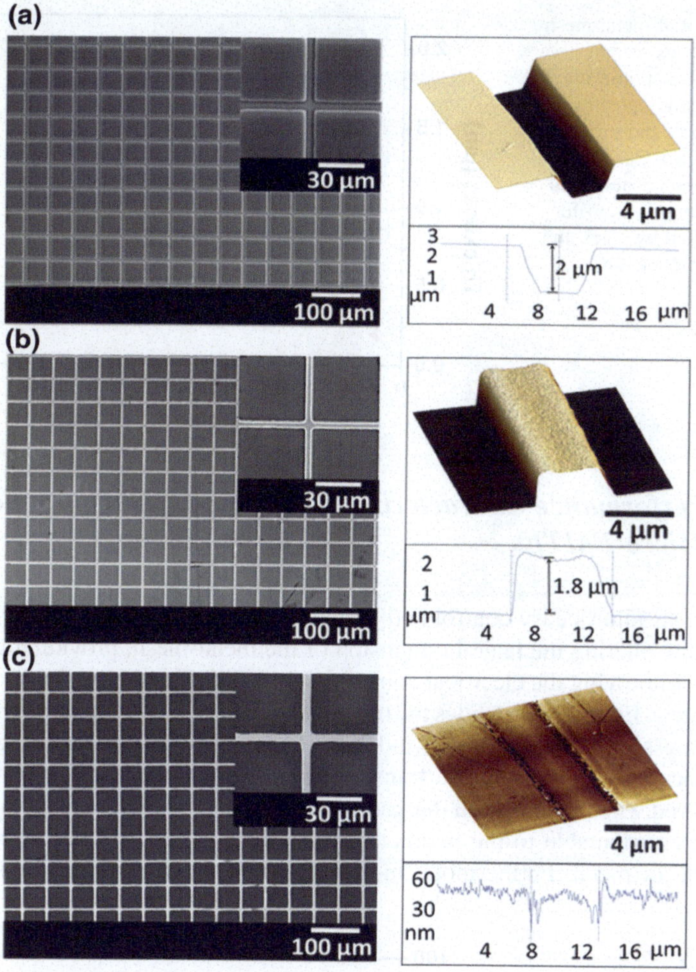

Fig. 3.2 Morphological characterizations by SEM (left) and AFM (right) of prototype 50 µm-pitch copper micro-EMTE at different fabrication stages of LEIT fabrication strategy: **a** as-developed mesh pattern in the photoresist; **b** copper mesh on the FTO glass substrate after removal of photoresist; **c** copper mesh transferred and fully embedded in a COC film

thinner metal meshes, including the interfacial friction and mechanical interlocking, could not counteract the adhesion force between the metal and FTO glass. Thus, the adhesion at the metal–FTO interface and the metal–COC interface as well as the geometric profile of the metal-mesh are crucial to reliable fabrication of EMTEs.

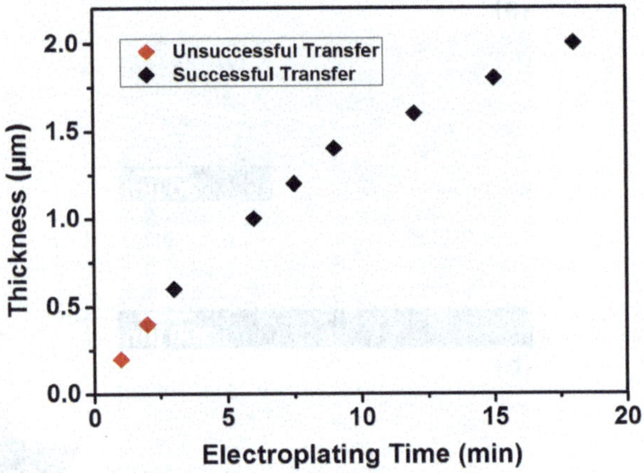

Fig. 3.3 Plot of metal-mesh thickness versus electrodeposition time at a constant electrodeposition current (5 mA) and substrate size (2 × 2 cm). Cases of successful and unsuccessful subsequent imprint transfer are represented by black and red colors respectively

3.3.2 Performance Characterizations of the LEIT Fabricated Micro-EMTEs

LEIT process allows easy control and variation of metal-mesh thickness while not considerably altering the lateral dimension of the metal-mesh, providing a feasible method of improving the electrical conductivity of EMTEs without sacrificing their transmittance. Figure 3.4 provides the transmittance of typical copper micro-EMTEs with thicknesses of 600 nm, 1 μm, and 2 μm in the 300–850 nm wavelength range. Only a marginal decrease in the transmittance over the measured spectral range was observed when metal-mesh thickness increased from 600 nm to 2 μm, and this decrease is attributable to the nonrectangular shape of the photoresist trench and overplating of metal. Furthermore, measurement of the normalized transmittance

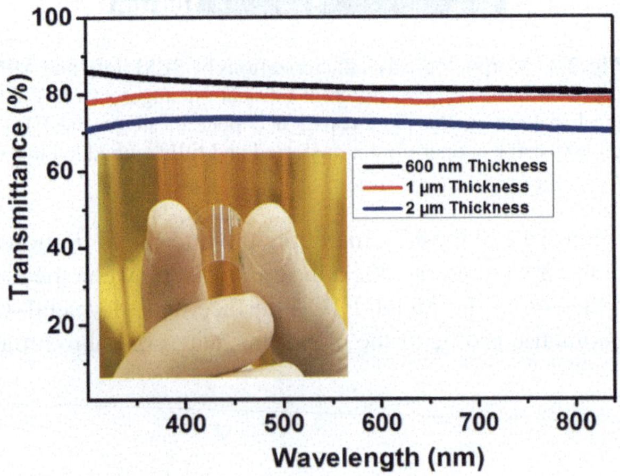

Fig. 3.4 UV–Vis spectra of the representative copper 50 μm pitch micro-EMTEs with thickness of 600 nm, 1 μm, and 2 μm

at different incident angles shows negligible angle dependence up to 60° from the normal direction (Fig. 3.12, Appendix).

On the other hand, sheet resistance of EMTEs can be substantially reduced when the metal thickness is increased, as displayed in Fig. 3.5a. An extremely low sheet resistance of 0.07 Ω/sq was observed for the 2 μm-thick copper micro-EMTE, and the transmittance at the 550 nm wavelength was still above 70%. To gain further insight into how metal thickness affects the overall performance of the EMTEs, a commonly used figure of merit (FoM), namely the ratio of electrical conductance to optical conductance (σ_{dc}/σ_{opt}), was calculated for all the micro-EMTEs displayed in Fig. 3.5a by using the following widely accepted expression [5–8].

$$\mathrm{FoM} = \frac{\sigma_{dc}}{\sigma_{opt}} = \frac{188.5}{\mathrm{R_s}\left(\frac{1}{\sqrt{\mathrm{T}}} - 1\right)} \tag{3.1}$$

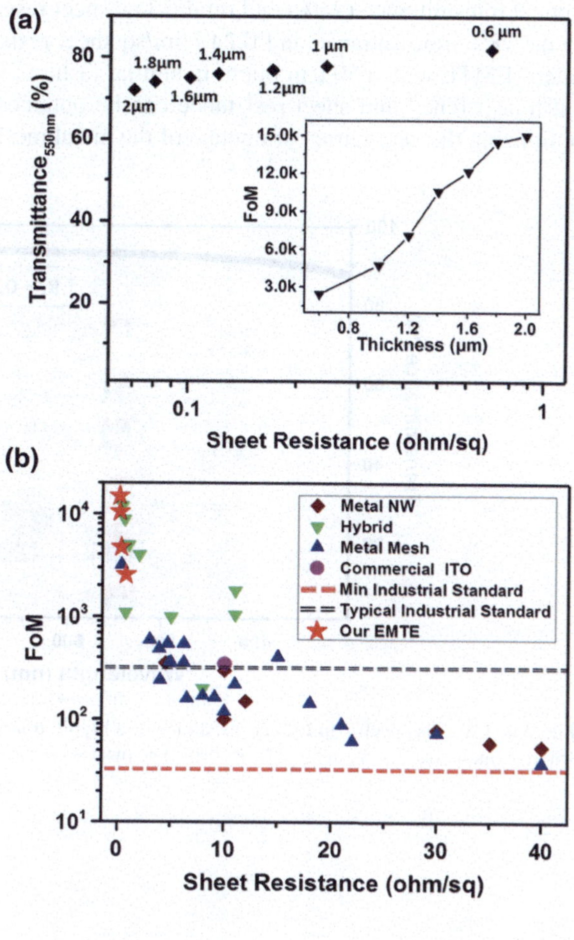

Fig. 3.5 Performance characterization of the prototype 50 μm-pitch copper micro-EMTEs. **a** Plot of transmittance versus sheet resistance with different mesh thickness, with calculated FoMs shown in the inset. **b** Comparison of the FoMs with other published TCs (metal NW [9–16], metal mesh [7, 8, 17–31] and hybrid [32–36]) and industrial standards [36]

where T is the optical transmittance at a wavelength of 550 nm and R_s is the sheet resistance. The inset of Fig. 3.5a provides the plot of the FoM as a function of metal thickness. The presented data indicate that the metal thickness has a considerable effect on the sheet resistance and hence on the value of the FoM by increasing the electrical conductivity of a thicker metal mesh without substantially sacrificing transmittance. Our micro-EMTEs achieved FoM of more than 1.5×10^4, which is among the highest such values ever reported. Figure 3.5b presents a comparison of our FoM with those of other transparent electrodes reported in recent studies. These data clearly indicate that our micro-EMTEs boast superior overall performance compared with most existing metal meshes, metal nanowire and hybrid TCs.

Increasing the gap size of the EMTE while maintaining same linewidth can improve the transmittance. Figure 3.6 displays the UV–Vis spectra and sheet resistance of a highly transparent copper micro-EMTE on a 5 × 5 cm large COC substrate with a pitch of 150 μm and with a linewidth and metal mesh thickness of 4 and 1 μm, respectively. Because of large line spacing, the sample exhibits high optical transmittance (94%) and retains low sheet resistance (0.93 Ω/sq), compared to the 78% transmittance and 0.24 ohm/sq sheet resistance of the previous copper micro-EMTE with a 50 μm pitch. In similar fashion, various combinations of optical transparency and sheet resistance can be obtained for various applications by optimizing the geometric parameters of the metal mesh.

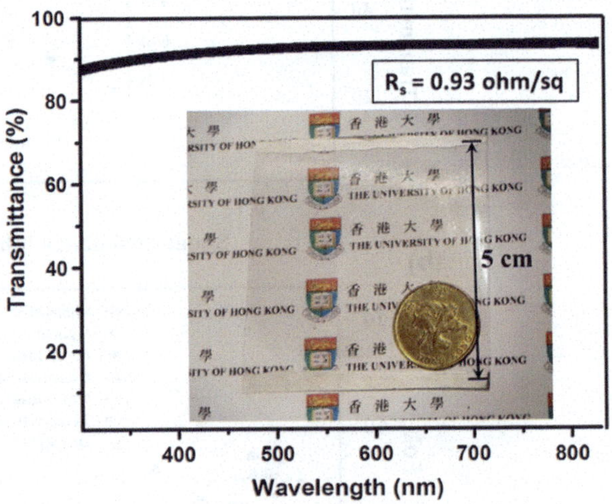

Fig. 3.6 UV–Vis spectra and sheet resistance of a highly transparent copper micro-EMTE with a 150 μm pitch on a 5 × 5 cm large COC film. The inset shows the optical image of the final structure

3.3.3 *Material Versatility of the LEIT Fabricated Micro-EMTEs*

Another crucial advantage of our transparent electrodes over random nanowire-based TCs is the wide range of usable materials. To demonstrate that our fabrication is versatile regarding material choice, micro-EMTEs of silver, gold, nickel, and zinc were fabricated on COC films. Figure 3.7 presents the UV–Vis transmittance spectra and sheet resistance of these micro-EMTEs. The transmittance spectra were nearly flat and featureless over the entire visible region, which is advantageous for display devices and solar cell applications. Silver mesh-, nickel mesh-, and zinc mesh-based EMTEs had similar metal thickness; therefore, all three samples had nearly the same transmittance (approximately 78% at a 550 nm wavelength), whereas the sheet resistances were 0.52, 1.40, and 1.02 Ω/sq, respectively. Because of varying metal thickness, the copper mesh- and gold mesh-based micro-EMTEs (approximately 600 and 2 μm, respectively) had transmittances of 82% and 72% and sheet resistances of 0.70 and 0.20 Ω/sq, respectively. The sheet resistance values of all the EMTEs were in accordance with the electrical resistivity values of the respective metals (Table 3.2, Appendix). The successful fabrication of these prototype micro-EMTEs validated the flexibility of material choice, thus fulfilling different requirements on the electrode's work function and chemical stability in different devices.

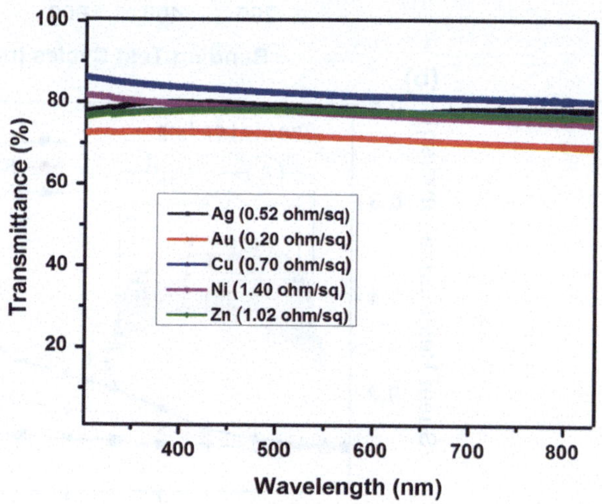

Fig. 3.7 UV–Vis spectra and sheet resistances of 50 μm-pitch flexible micro-EMTEs fabricated with various metals showing versatile material choices

3.3.4 Mechanical Stability of the LEIT Fabricated Micro-EMTEs

In addition to enhancing the electrical and optical performance of the EMTE, the embedded nature of the metal mesh greatly improves its adhesion with the substrate and enhances its stability under bending, heating, and chemical attack. Figure 3.8a, b provide the test results of mechanical stability on copper micro-EMTEs under cyclic

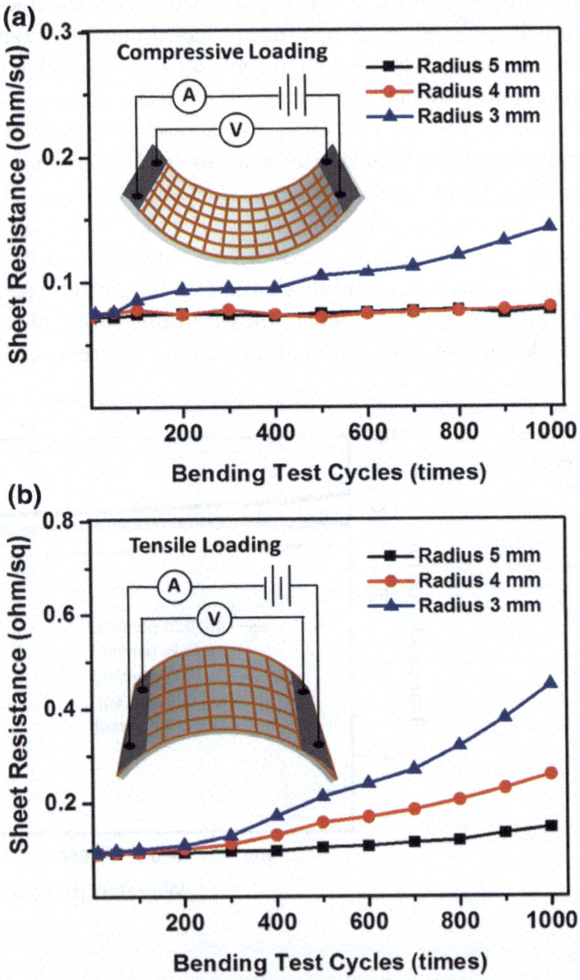

Fig. 3.8 Mechanical stability of the 50 μm-pitch flexible copper micro-EMTEs. **a** Plot of variations in sheet resistance versus the number of cycles of repeated bending (compressive loading) to radii of 5 mm, 4 mm, and 3 mm. **b** Plot of variations in sheet resistance versus the number of cycles of repeated bending (tensile loading) to radii of 5 mm, 4 mm, and 3 mm

tensile and compressive bending stress. Figure 3.8a presents the variation in sheet resistance as a function of the number of cycles for repeated compressive bending to radii of 5, 4, and 3 mm. The results clearly indicate that for 5 and 4 mm bending radii, no noticeable change in sheet resistance (0.07 Ω/sq) occurs for up to 1000 bending cycles. For a 3 mm bending radius, the change in sheet resistance is within 100% of its original value (from 0.07 to 0.13 Ω/sq). This remarkable copper mesh stability is attributable to its embedded nature. Similarly, for tensile loading, changes in sheet resistance versus the number of repeated bending cycles are displayed in Fig. 3.8b, revealing that for 1000 repeated bendings to 5, 4, and 3 mm radii, the sheet resistances varied by approximately 30% (from 0.10 to 0.13 Ω/sq), 150% (from 0.10 to 0.25 Ω/sq), and 350% (from 0.10 to 0.45 ^{O}/sq), respectively. These higher increases in sheet resistances presumably arose because of metal-mesh cracking (Fig. 3.13, Appendix) under high tensile stress, indicating that the metal mesh is more vulnerable to failure under tensile stress than compressive stress. Further studies are required to investigate the failure mechanisms of the metal mesh under various bending loads and their effects on the performance of the EMTEs and derived devices. The mechanical stability of the micro-EMTEs was also confirmed using a repeated adhesive tape test. The sheet resistance of a typical copper micro-EMTE was measured after every 10 peeling tests using polypropylene tape with acrylic adhesive and was found to be unchanged after 100 cycles (Fig. 3.14, Appendix), confirming that the strong adhesion is due to the mechanical anchoring of the embedded metal mesh in the substrate.

3.3.5 Environmental Stability of the LEIT Fabricated Micro-EMTEs

The environmental stability of the as-fabricated copper micro-EMTEs was also evaluated by dipping them in DI water and isopropyl alcohol and exposing them to high humidity and high temperature conditions (60 °C, 85% relative humidity). Figure 3.9 shows that after 24 h, the sheet resistances and morphological structures of the micro-EMTEs remained unchanged. This superior stability of the copper micro-EMTEs was further confirmed by performing the EDS analysis on both as-fabricated copper micro-EMTE (Fig. 3.15, Appendix) and the treated copper micro-EMTEs (Figs. 3.16, 3.17 and 3.18, Appendix). The favorable environmental stability demonstrated in these tests can also be attributed to the embedded nature of the EMTEs and the chemical stability of the COC films that isolate and protect the metal-mesh.

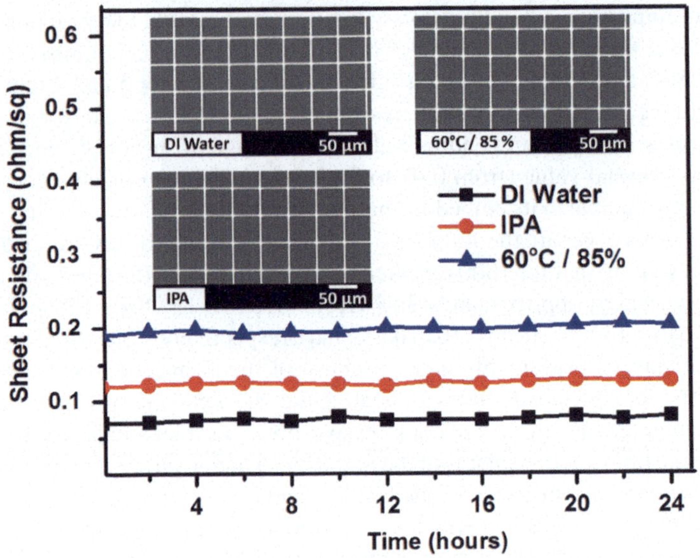

Fig. 3.9 Variations in sheet resistance of 50 μm-pitch flexible copper micro-EMTEs during the chemical and environmental stability tests. Inset: SEM images after the tests

Appendix: Micro Embedded Metal-mesh Transparent Electrodes

See Figs. 3.10, 3.11, 3.12, 3.13, 3.14, 3.15, 3.16, 3.17 and 3.18; Tables 3.1 and 3.2.

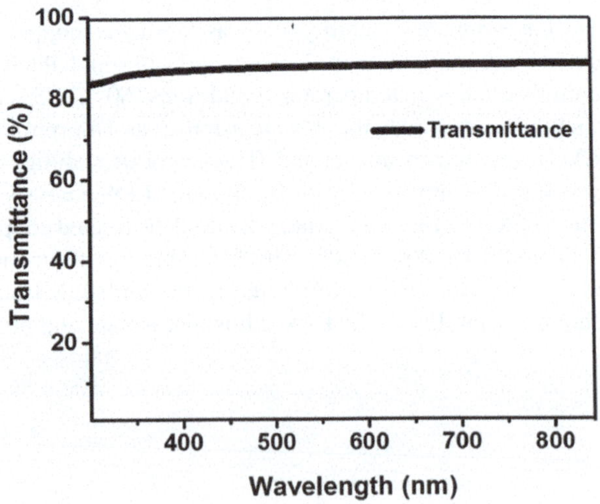

Fig. 3.10 Optical transparency of blank COC film in UV and visible wavelength range

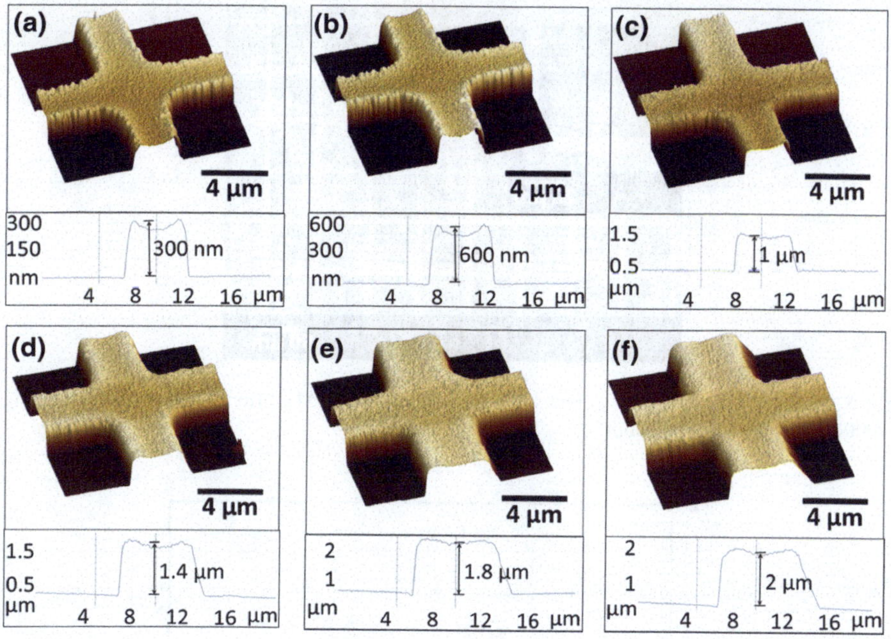

Fig. 3.11 AFM images showing height profiles and sections of electroplated copper meshes (p = 50 µm) on FTO glass (substrate size: 2 cm × 2 cm), fabricated at constant deposition current of 5 mA. **a** Mesh thickness ~300 nm, electroplated for 1.5 min; **b** mesh thickness ~600 nm, electroplated for 3 min; **c** mesh thickness ~1 µm, electroplated for 6 min; **d** mesh thickness ~1.4 µm, electroplated for 9 min; **e** mesh thickness ~1.8 µm, electroplated for 15 min; **f** mesh thickness ~2 µm, electroplated for 18 min

Fig. 3.12 The measured total transmission spectra of a typical micro-EMTE sample at different incident angles

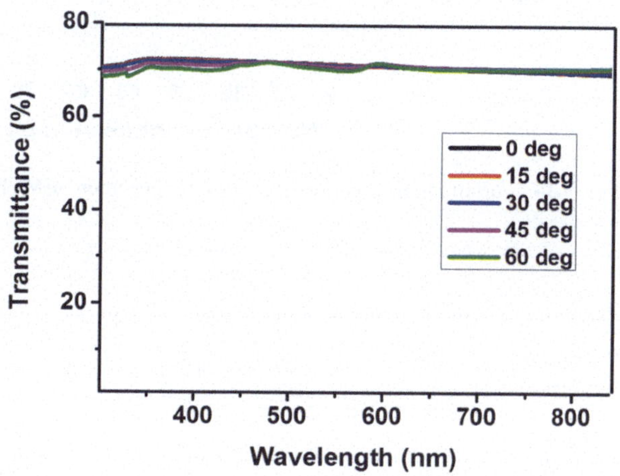

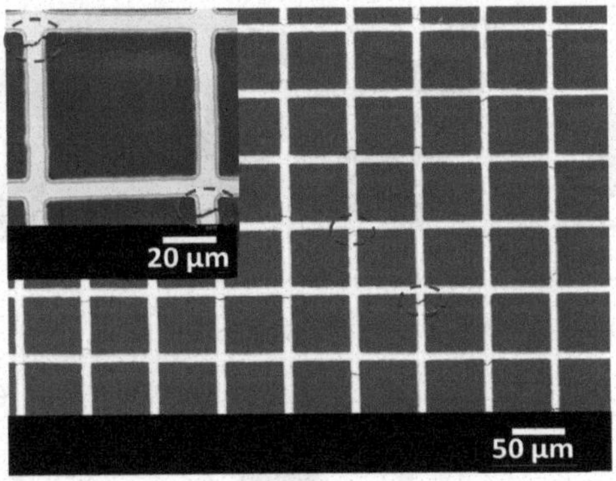

Fig. 3.13 SEM images showing cracks in the copper micro-EMTE after repeated tensile bending (1000 cycles) to radius of 3 mm

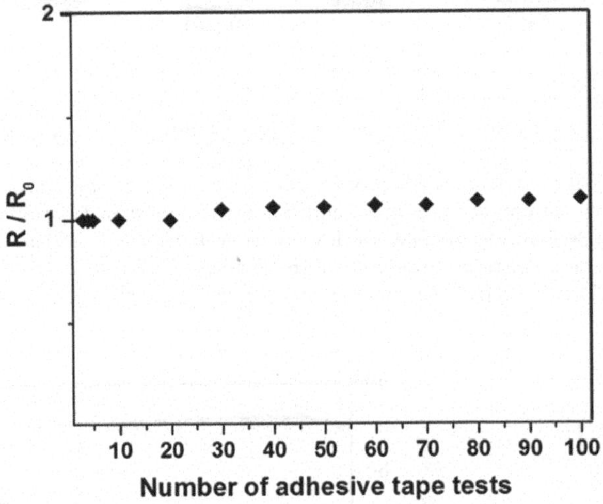

Fig. 3.14 Variations in sheet resistance of copper micro-EMTE after the repeated adhesive tape tests

(a)

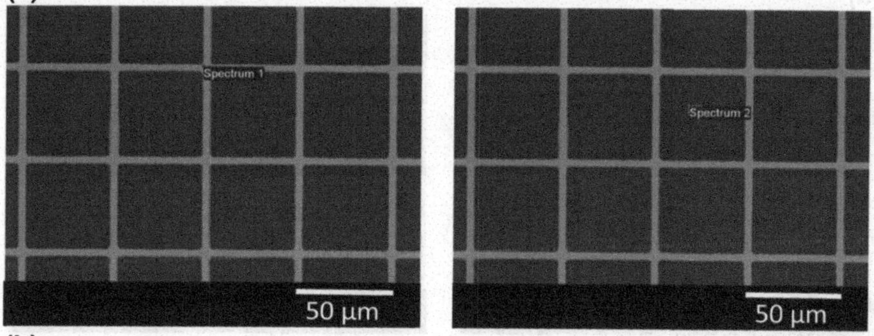

(b)

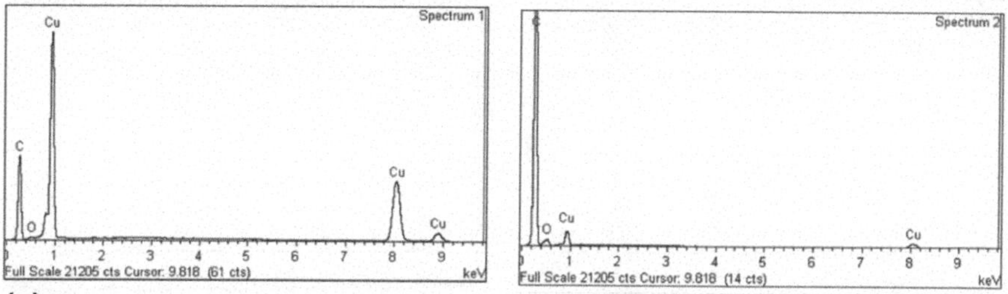

(c)

Element	App (Conc.)	Intensity	Weight%	Weight% (Sigma)	Atomic%
C K	133.51	0.5190	52.99	0.25	84.93
O K	2.43	0.5434	0.92	0.11	1.11
Cu K	187.21	0.8368	46.09	0.25	13.96
Totals			100.00		

Element	App (Conc.)	Intensity	Weight%	Weight% (Sigma)	Atomic%
C K	440.08	1.5036	88.38	0.32	93.98
O K	6.79	0.3327	6.17	0.31	4.92
Cu K	13.35	0.7388	5.46	0.12	1.10
Totals			100.00		

(d)

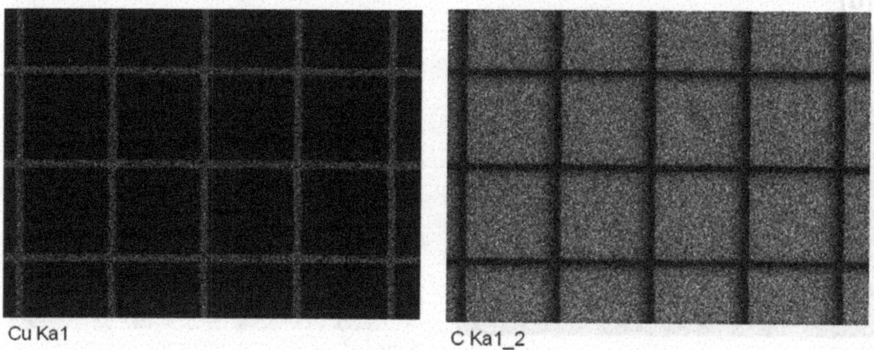

Fig. 3.15 SEM-EDS analysis for a typical as-fabricated copper micro-EMTE sample at copper mesh (left) and COC film (right) before chemical stability tests **a** SEM micrographs **b** EDS spectrum of the corresponding boxed area in (a) **c** elemental quantification tables at the corresponding boxed area in (a) **d** elemental maps; copper (left) and carbon (right)

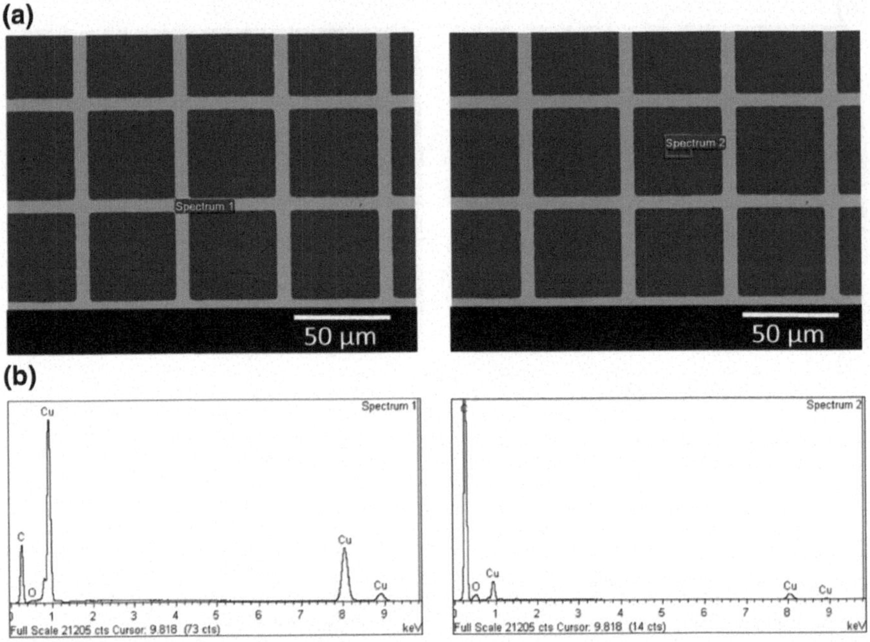

(c)

Element	App (Conc.)	Intensity	Weight%	Weight% (Sigma)	Atomic%
C K	109.16	0.4893	49.21	0.27	83.01
O K	2.19	0.5718	0.85	0.11	1.07
Cu K	191.78	0.8471	49.94	0.27	15.92
Totals			100.00		

Element	App (Conc.)	Intensity	Weight%	Weight% (Sigma)	Atomic%
C K	431.19	1.3897	86.78	0.29	93.70
O K	7.23	0.3406	5.94	0.29	4.81
Cu K	19.35	0.7429	7.28	0.13	1.49
Totals			100.00		

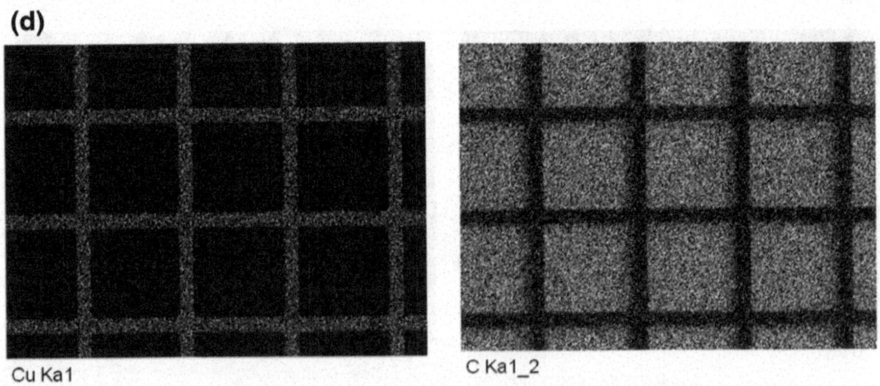

Fig. 3.16 SEM-EDS analysis for a typical as-fabricated copper micro-EMTE sample at copper mesh (left) and COC film (right) after dipping in IPA for 24 h **a** SEM micrographs **b** EDS spectrum of the corresponding boxed area in (a) **c** elemental quantification tables at the corresponding boxed area in (a) **d** elemental maps; copper (left) and carbon (right)

(a)

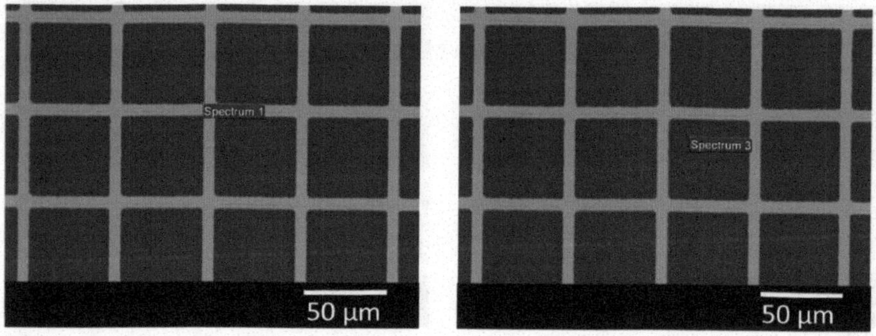

(b)

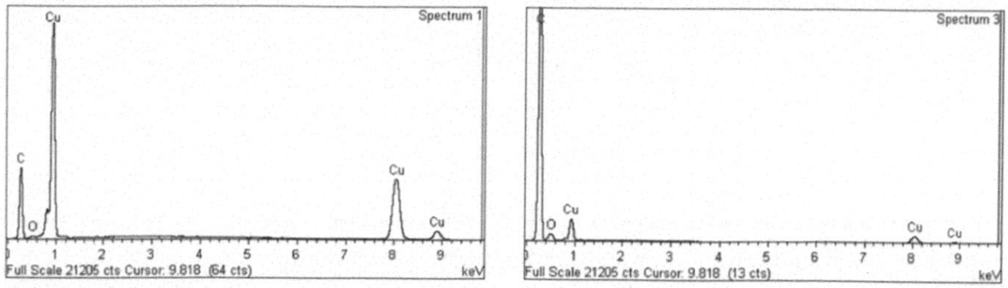

(c)

Element	App (Conc.)	Intensity	Weight%	Weight% (Sigma)	Atomic%
C K	114.23	0.4941	49.79	0.26	83.25
O K	2.48	0.5674	0.94	0.11	1.18
Cu K	193.37	0.8453	49.27	0.26	15.57
Totals			100.00		

Element	App (Conc.)	Intensity	Weight%	Weight% (Sigma)	Atomic%
C K	430.91	1.3836	86.58	0.29	93.57
O K	7.46	0.3415	6.07	0.28	4.93
Cu K	19.66	0.7431	7.35	0.13	1.50
Totals			100.00		

(d)

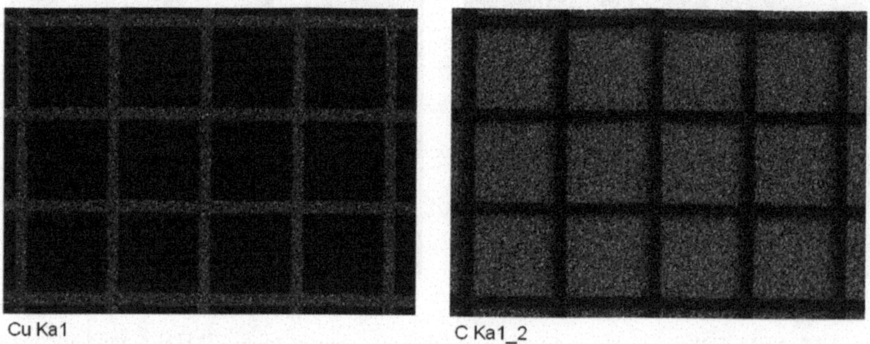

Fig. 3.17 SEM-EDS analysis for a typical as-fabricated copper micro-EMTE sample at copper mesh (left) and COC film (right) after exposing them to high humidity and high temperature conditions (60 °C, 85% relative humidity) for 24 h **a** SEM micrographs **b** EDS spectrum of the corresponding boxed area in (a) **c** elemental quantification tables at the corresponding boxed area in (a) **d** elemental maps; copper (left) and carbon (right)

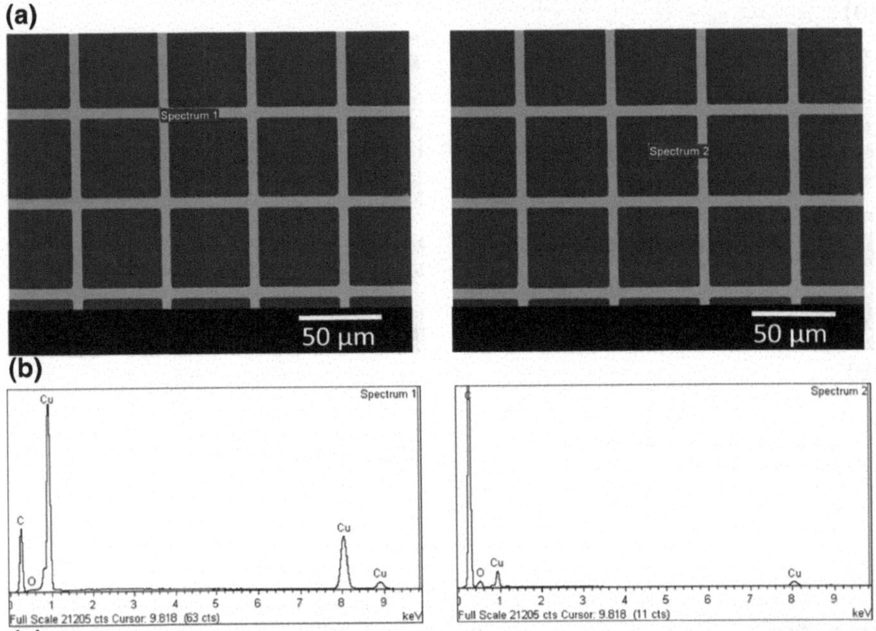

(c)

Element	App (Conc.)	Intensity	Weight%	Weight% (Sigma)	Atomic%
C K	117.74	0.4992	50.49	0.26	83.64
O K	2.43	0.5621	0.92	0.11	1.15
Cu K	191.44	0.8435	48.59	0.26	15.21
Totals			100.00		

Element	App (Conc.)	Intensity	Weight%	Weight% (Sigma)	Atomic%
C K	454.23	1.4598	87.52	0.30	93.61
O K	7.70	0.3367	6.43	0.30	5.16
Cu K	15.90	0.7402	6.04	0.12	1.22
Totals			100.00		

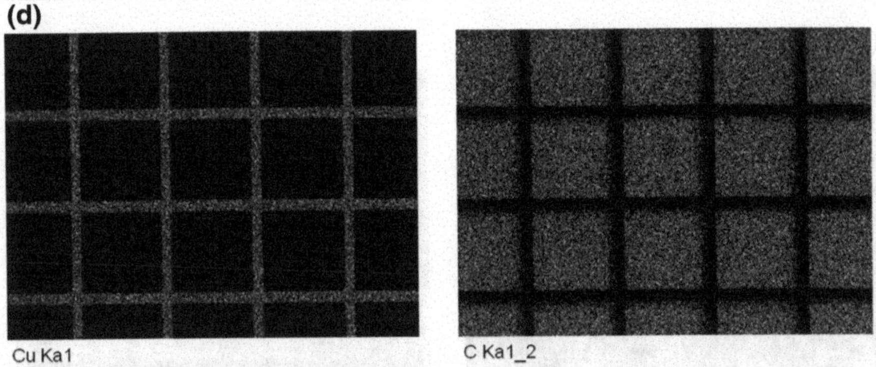

Fig. 3.18 SEM-EDS analysis for a typical as-fabricated copper micro-EMTE sample at copper mesh (left) and COC film (right) after dipping in DI water for 24 h **a** SEM micrographs **b** EDS spectrum of the corresponding boxed area in (a) **c** elemental quantification tables at the corresponding boxed area in (a) **d** elemental maps; copper (left) and carbon (right)

Table 3.1 Properties of the COC film

Property	Value
Film thickness (μm)	100
Glass transition temperature (°C)	78
Coefficient of linear thermal expansion (K^{-1})	0.7×10^{-4}
Young's modulus (MPa)	2600
Water Absorption @24 h immersion at 23 °C (%)	<0.01
Water contact angle (°)	93 ± 1.5

Table 3.2 Performance characterization of micro-EMTEs of various metals (p = 50 μm). For comparison the resistivity of each metal are listed. The difference in the sheet resistance of micro-EMTEs is due to the resistivity (material property) difference and metal thickness. For all the EMTEs, values of the sheet resistances are in accordance with the electrical resistivity values

Material	Resistivity (Ω-m) @ 25 °C	Mesh-thickness (μm)	Sheet resistance (Ω-m)	Transmittance (%) @ 550 nm
Copper	1.68×10^{-8}	0.65	0.70	81.84
Silver	1.59×10^{-8}	0.85	0.52	78.93
Nickel	6.99×10^{-8}	1.20	1.40	77.78
Gold	2.44×10^{-8}	1.90	0.20	71.65
Zinc	5.90×10^{-8}	1.15	1.02	78.04

References

1. Jang T-H, Zhang C, Youn H-S, Zhou J, Guo LJ (2015) IEEE transaction on antenna and propagation, (in review)
2. Kim DS, Khan A, Rahman K, Khan S, Kim HC, Choi KH (2011) Drop-on-demand direct printing of colloidal copper nanoparticles by electrohydrodynamic atomization. Mater Manuf Process 26(9):1196–1201. https://doi.org/10.1080/10426914.2011.551956
3. Khalid R, Arshad K, Nauman Malik M, Jeongdai J, Kyung-Hyun C (2012) Fine-resolution patterning of copper nanoparticles through electrohydrodynamic jet printing. J Micromech Microeng 22(6):065012
4. Kim DS, Rahman K, Khan A, Choi KH (2012) Direct fabrication of copper nanoparticle patterns through electrohydrodynamic printing in cone-jet mode. Mater Manuf Process 27(12):1295–1299. https://doi.org/10.1080/10426914.2012.663121
5. van de Groep J, Spinelli P, Polman A (2012) Transparent conducting silver nanowire networks. Nano Lett 12(6):3138–3144. https://doi.org/10.1021/nl301045a
6. Vosgueritchian M, Lipomi DJ, Bao Z (2012) Highly conductive and transparent pedot: pss films with a fluorosurfactant for stretchable and flexible transparent electrodes. Adv Funct Mater 22(2):421–428. https://doi.org/10.1002/adfm.201101775
7. Han B, Pei K, Huang Y, Zhang X, Rong Q, Lin Q, Guo Y, Sun T, Guo C, Carnahan D, Giersig M, Wang Y, Gao J, Ren Z, Kempa K (2014) Uniform self-forming metallic network as a high-performance transparent conductive electrode. Adv Mater 26(6):873–877. https://doi.org/10.1002/adma.201302950
8. Kim H-J, Lee S-H, Lee J, Lee E-S, Choi J-H, Jung J-H, Jung J-Y, Choi* D-G, (2014) High-durable AgNi nanomesh film for a transparent conducting electrode. Small 10(18):3767–3774. https://doi.org/10.1002/smll.201400911

9. Rathmell AR, Bergin SM, Hua Y-L, Li Z-Y, Wiley BJ (2010) The growth mechanism of copper nanowires and their properties in flexible, transparent conducting films. Adv Mater 22(32):3558–3563. https://doi.org/10.1002/adma.201000775
10. Wu H, Hu L, Rowell MW, Kong D, Cha JJ, McDonough JR, Zhu J, Yang Y, McGehee MD, Cui Y (2010) Electrospun metal nanofiber webs as high-performance transparent electrode. Nano Lett 10(10):4242–4248. https://doi.org/10.1021/nl102725k
11. Zhu R, Chung C-H, Cha KC, Yang W, Zheng YB, Zhou H, Song T-B, Chen C-C, Weiss PS, Li G, Yang Y (2011) Fused silver nanowires with metal oxide nanoparticles and organic polymers for highly transparent conductors. ACS Nano 5(12):9877–9882. https://doi.org/10.1021/nn203576v
12. Lee J, Lee P, Lee H, Lee D, Lee SS, Ko SH (2012) Very long Ag nanowire synthesis and its application in a highly transparent, conductive and flexible metal electrode touch panel. Nanoscale 4(20):6408–6414. https://doi.org/10.1039/C2NR31254A
13. Tokuno T, Nogi M, Jiu J, Sugahara T, Suganuma K (2012) Transparent electrodes fabricated via the self-assembly of silver nanowires using a bubble template. Langmuir 28(25):9298–9302. https://doi.org/10.1021/la300961m
14. Zhang D, Wang R, Wen M, Weng D, Cui X, Sun J, Li H, Lu Y (2012) Synthesis of ultralong copper nanowires for high-performance transparent electrodes. J Am Chem Soc 134(35):14283–14286. https://doi.org/10.1021/ja3050184
15. Guo H, Lin N, Chen Y, Wang Z, Xie Q, Zheng T, Gao N, Li S, Kang J, Cai D, Peng D-L (2013) Copper nanowires as fully transparent conductive electrodes. Sci Rep 3:2323. https://doi.org/10.1038/srep02323, http://www.nature.com/articles/srep02323#supplementary-information
16. Wang J, Yan C, Kang W, Lee PS (2014) High-efficiency transfer of percolating nanowire films for stretchable and transparent photodetectors. Nanoscale 6(18):10734–10739. https://doi.org/10.1039/C4NR02462A
17. Kang MG, Guo LJ (2007) Nanoimprinted semitransparent metal electrodes and their application in organic light-emitting diodes. Adv Mater 19(10):1391–1396. https://doi.org/10.1002/adma.200700134
18. Kang M-G, Kim M-S, Kim J, Guo LJ (2008) Organic solar cells using nanoimprinted transparent metal electrodes. Adv Mater 20(23):4408–4413. https://doi.org/10.1002/adma.200800750
19. Zou J, Yip H-L, Hau SK, Jen AK-Y (2010) Metal grid/conducting polymer hybrid transparent electrode for inverted polymer solar cells. Appl Phys Lett 96(20):203301. https://doi.org/10.1063/1.3394679
20. Hong S, Yeo J, Kim G, Kim D, Lee H, Kwon J, Lee H, Lee P, Ko SH (2013) Nonvacuum, maskless fabrication of a flexible metal grid transparent conductor by low-temperature selective laser sintering of nanoparticle ink. ACS Nano 7(6):5024–5031. https://doi.org/10.1021/nn400432z
21. Wu H, Kong D, Ruan Z, Hsu P-C, Wang S, Yu Z, Carney TJ, Hu L, Fan S, Cui Y (2013) A transparent electrode based on a metal nanotrough network. Nat. Nano 8(6):421–425. https://doi.org/10.1038/nnano.2013.84, http://www.nature.com/nnano/journal/v8/n6/abs/nnano.2013.84.html#supplementary-information
22. Yonghee J, Jihoon K, Doyoung B (2013) Invisible metal-grid transparent electrode prepared by electrohydrodynamic (EHD) jet printing. J Phys D: Appl Phys 46(15):155103
23. Guo CF, Sun T (2014) Liu Q.; Suo Z.; Ren Z. Highly stretchable and transparent nanomesh electrodes made by grain boundary lithography. Nat Commun 5. https://doi.org/10.1038/ncomms4121
24. Gupta R, Rao KDM, Srivastava K, Kumar A, Kiruthika S, Kulkarni GU (2014) Spray coating of crack templates for the fabrication of transparent conductors and heaters on flat and curved surfaces. ACS Appl Mater Interfaces 6(16):13688–13696. https://doi.org/10.1021/am503154z
25. Iwahashi T, Yang R, Okabe N, Sakurai J, Lin J, Matsunaga D (2014) Nanoimprint-assisted fabrication of high haze metal mesh electrode for solar cells. Appl Phys Lett 105(22):223901. https://doi.org/10.1063/1.4903061

26. Kiruthika S, Gupta R, Rao KDM, Chakraborty S, Padmavathy N, Kulkarni GU (2014) Large area solution processed transparent conducting electrode based on highly interconnected Cu wire network. J Mater Chem C 2(11):2089–2094. https://doi.org/10.1039/C3TC32167C
27. Kiruthika S, Rao KDM, Ankush K, Ritu G, Kulkarni GU (2014) Metal wire network based transparent conducting electrodes fabricated using interconnected crackled layer as template. Mater Res Express 1(2):026301
28. Park JH, Lee DY, Kim Y-H, Kim JK, Lee JH, Park JH, Lee T-W, Cho JH (2014) Flexible and Transparent Metallic Grid Electrodes Prepared by Evaporative Assembly. ACS Appl Mater Interfaces 6(15):12380–12387. https://doi.org/10.1021/am502233y
29. Zhou L, Xiang H-Y, Shen S, Li Y-Q, Chen J-D, Xie H-J, Goldthorpe IA, Chen L-S, Lee S-T, Tang J-X (2014) High-performance flexible organic light-emitting diodes using embedded silver network transparent electrodes. ACS Nano 8(12):12796–12805. https://doi.org/10.1021/nn506034g
30. Choi H-J, Choo S, Jung P-H, Shin J-H, Kim Y-D, Lee H (2015) Uniformly embedded silver nanomesh as highly bendable transparent conducting electrode. Nanotechnology 26(5):055305
31. Kang J, Jang Y, Kim Y, Cho S-H, Suhr J, Hong BH, Choi J-B, Byun D (2015) An Ag-grid/graphene hybrid structure for large-scale, transparent, flexible heaters. Nanoscale 7(15):6567–6573. https://doi.org/10.1039/C4NR06984F
32. Hsu P-C, Wang S, Wu H, Narasimhan VK, Kong D, Ryoung Lee H, Cui Y (2013) Performance enhancement of metal nanowire transparent conducting electrodes by mesoscale metal wires. Nat Commun 4. https://doi.org/10.1038/ncomms3522
33. Ok JG, Kwak MK, Huard CM, Youn HS, Guo LJ (2013) Photo-roll lithography (PRL) for continuous and scalable patterning with application in flexible electronics. Adv Mater 25(45):6554–6561. https://doi.org/10.1002/adma.201303514
34. An BW, Hyun BG, Kim S-Y, Kim M, Lee M-S, Lee K, Koo JB, Chu HY, Bae B-S, Park J-U (2014) Stretchable and transparent electrodes using hybrid structures of graphene-metal nanotrough networks with high performances and ultimate uniformity. Nano Lett 14(11):6322–6328. https://doi.org/10.1021/nl502755y
35. Bao W, Wan J, Han X, Cai X, Zhu H, Kim D, Ma D, Xu Y, Munday JN, Drew HD, Fuhrer MS, Hu L (2014) Approaching the limits of transparency and conductivity in graphitic materials through lithium intercalation. Nat Commun 5. https://doi.org/10.1038/ncomms5224
36. Gao T, Li Z, Huang P-S, Shenoy GJ, Parobek D, Tan S, Lee J-K, Liu H, Leu PW (2015) Hierarchical Graphene/Metal grid structures for stable. Flexible transparent conductors. ACS Nano 9(5):5440–5446. https://doi.org/10.1021/acsnano.5b01243

Chapter 4
Micro Embedded Metal-mesh Transparent Electrodes (Micro-EMTEs) Fabricated by TEIT Strategy

Abstract This chapter discusses an improved technique for the fabrication of micro-EMTEs is developed by eliminating the mandatory mesh patterning step from the unit-production cycle of LEIT. In this TEIT strategy, a reusable SU-8 based electrodeposition template with micro-mesh features is engaged to simplify the fabrication of micro-EMTEs. Based on this novel technique, prototype flexible micro-EMTEs with nickel mesh of various pitches are fabricated on COC films, demonstrating excellent electrical and optical performances. The stability of SU-8 templates is also tested; no degradation in the performance and morphology of the template is observed after utilizing it for 20 repeated fabrication cycles. These excellent results make SU-8 a promising candidate material for reusable electrodeposition templates for large-scale manufacturing of micro-EMTEs.

4.1 Introduction to TEIT Fabrication Strategy for Micro-EMTEs

As discussed in Chap. 3, EMTE's structure and the LEIT fabrication strategy address most of the key shortcomings of metal mesh based TCs by eliminating the expensive vacuum-based metal deposition, improving the surface topography of metal-mesh, and enhancing adhesion between metal mesh and the substrate. Though LEIT is a cost-effective approach for fabrication of EMTEs, yet it comprises a mandatory lithography step in making each sample which limits its suitability for high-throughput and large-volume industrial production. In this chapter, we demonstrate an improved technique based on templated electrodeposition for fabrication of micro-EMTEs by eliminating the obligatory lithography step from the unit-production cycle of LEIT. In this TEIT strategy, a reusable template with micro-mesh features is employed to simplify the fabrication of micro-EMTEs.

The fabrication procedure is defined in Fig. 4.1. In a usual fabrication practice, first, a 500 nm SU-8 photoresist layer is spin coated on a cleaned ITO glass substrate (Fig. 4.1a). Then, the micro-mesh patterns are prepared in SU-8 film by photolithography, and employed as repetitive electrodeposition template (Fig. 4.1b). Next, metal is deposited in the trenches of the mesh patterns through electrodeposition (Fig. 4.1c).

A. Khan, *Novel Embedded Metal-mesh Transparent Electrodes*, Springer Theses,
https://doi.org/10.1007/978-981-15-2918-4_4

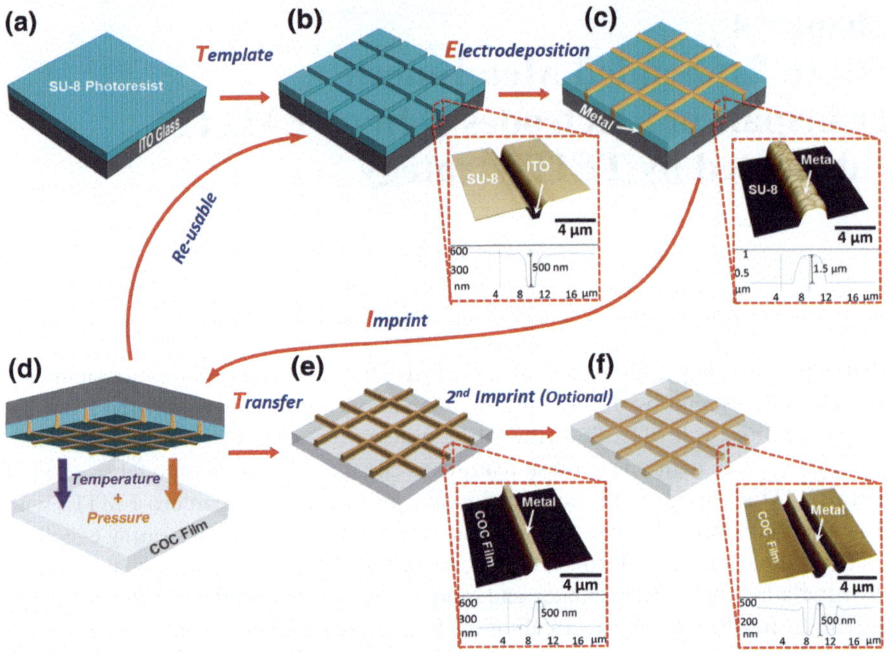

Fig. 4.1 Schematic illustrations of TEIT fabrication strategy for micro-EMTEs. **a** SU-8 photoresist layer on ITO glass substrate formed by spin coating. **b** Mesh patterns formed in SU-8 film by photolithography. **c** Electrodeposition of metal inside the resist trenches to form a uniform metal-mesh. **d** Heating and pressing the metal-mesh into a COC film. **e** Peeling off the COC film with the metal-mesh transferred in a partial embedded form. **f** Second round of thermal imprinting using a featureless mold to fully-embed the metal-mesh into COC film

With optimized overplated metal thickness, the COC film is heated to above its glass transition temperature and pressed into the metal-mesh (Fig. 4.1d). Finally, after cooling the stack and separating the COC film from the SU-8 template, the metal-mesh is transferred in partial embedded form (Fig. 4.1e). Fully-embedded micro-EMTE is then accomplished by second round of thermal imprinting using a featureless mold (Fig. 4.1f). The SU-8 template is then ready for the next production cycle eliminating the laborious mesh-template patterning step, making the TEIT fabrication strategy more facile as compared to LEIT.

4.2 Experimental Details

4.2.1 Fabrication of Reusable Electrodeposition Templates

First, ITO glass substrates (approximately 10 Ω/sq, South China Xiang S&T, China) were cleaned with a cotton swab and liquid detergent, rinsed thoroughly with another cotton swab and DI water. ITO glass substrate was then put in isopropanol and DI water for ultrasonication for 30 s before being dried with compressed air. Afterwards, substrates were further cleaned by oxygen plasma for 10 min. Next, the negative photoresist SU-8 2000 (Microchem, USA) was diluted to 10% solid contents concentration using a thinner solution cyclopentanone (Microchem, USA). Then, the diluted SU-8 photoresist was spin coated on ITO glass substrate at 2000 rpm for 60 s to reach a film thickness of 500 nm. The film was then prebaked on a hotplate at 110 °C for 60 s. Next, the SU-8 film was exposed using a URE-2000/35 UV mask aligner (Chinese Academy of Sciences) for an exposure dose of 60 mJ/cm^2. Afterwards, the film was post-baked on a hotplate at 110 °C for 60 s. The SU-8 film was then developed in propylene glycol monomethyl ether acetate (PGMEA) developer (Microchem, USA) for 50 s. Thereafter, the sample was rinsed in DI water and blow-dried with compressed air. Finally, the template was hard-baked at 200 °C for 90 min to promote the adhesion between SU-8 film and ITO glass substrate.

4.2.2 Fabrication of Flexible Micro-EMTEs by TEIT Strategy

Once the reusable template with micro-patterns in SU-8 film was ready, the electrodeposition was performed using commercial nickel plating solution (Caswell, USA). A Keithley 2400 SourceMeter was used to supply a constant 5 mA current to a two-electrode electrodeposition setup with the SU-8 based electrodeposition template as the working electrode and metal bar as the counter electrode. Electrodeposition time was optimized to reach the desired thickness of deposited metal for successful transfer. After the electrodeposition was completed, the samples were thoroughly rinsed with DI water, blow-dried with compressed air. The metal-mesh was then transferred to a 100-μm-thick COC film (Grade 6017) by thermal imprint using the same home-built setup described in Chap. 3 (Sect. 3.2). During the thermal imprint step, the plates were heated to 210 °C and an imprinting pressure of 15 MPa was applied, holding it for 5 min. The heated platens were then cooled to the demolding temperature of 50 °C. Finally, the COC film was peeled from the template, with the metal-mesh partially embedded in the COC film and template was ready for the next production cycle. For fully-embedded micro-EMTEs, the COC film with partial embedded metal-mesh was heated again to 210 °C and pressed between two featureless silicon wafers so that the metal-mesh pushed into the thermally softened COC film. After cooling and releasing the pressure, the fully-embedded micro-EMTEs were ready for further device fabrication. The morphology of the samples was characterized using an S-3400N SEM (Hitachi, Japan) and a Multimode-8 AFM (Bruker, USA). The sheet resistance of the EMTE samples was measured using the four-probe method to eliminate contact resistance following the same protocol as described in Chap. 3 (Sect. 3.2). Similarly, optical transmission spectra were recorded using a Lambda 25 UV–Vis spectrometer (Perkin Elmer, USA). All transmittance values presented in this chapter are also normalized to the absolute transmittance through the bare COC substrate.

4.3 Experimental Results

4.3.1 *Morphological Characterizations of the TEIT Fabricated Micro-EMTEs*

As demonstration, a sample micro-EMTE with nickel mesh on COC film was fabricated by TEIT strategy. Figure 4.2a–d display the morphological characterization of the micro-EMTE at different stages of the fabrication. Figure 4.2a shows the SEM and AFM images of the trench in the SU-8 film produced by photolithography (Fig. 4.1b). On this sample, the photoresist trench had a 50 μm pitch, and its trench-width and -depth were approximately 2 μm and 500 nm, respectively. Figure 4.2b demonstrates the electroplated nickel mesh on the SU-8 template (Fig. 4.1c). The nickel mesh was grown with a 5 mA current over a 2 × 2 cm area. As obvious from the pictures, the nickel mesh had linewidth and thickness of approximately 4 and 2 μm, respectively. Figure 4.2c shows the nickel mesh finally transferred onto COC film (Fig. 4.1e). The AFM characterization reveals that the surface of nickel mesh is 500 nm (comparable to the trench-depth of the SU-8 template) above the level of COC film, confirming its partial embedded shape. Figure 4.2d displays the final fully-embedded micro-EMTE structure (Fig. 4.1f).

4.3.2 *Performance Characterizations of the TEIT Fabricated Micro-EMTEs*

Adjusting the pitch of the EMTE while maintaining the linewidth and thickness constant, its optical and electrical performance can easily be tuned. This way, different combinations of optical transparency and sheet resistance can be obtained for the required applications. Prototype micro-EMTEs with nickel mesh of varying pitches i.e. 25, 50, 75 and 100 μm were fabricated on COC films by TEIT process. For illustration, Fig. 4.3 displays the photograph of the 75 μm pitch nickel micro-EMTE placed in front of a building, illustrating its high transparency. The SEM images of these sample nickel mesh micro-EMTEs are presented in Fig. 4.4a–d, demonstrating

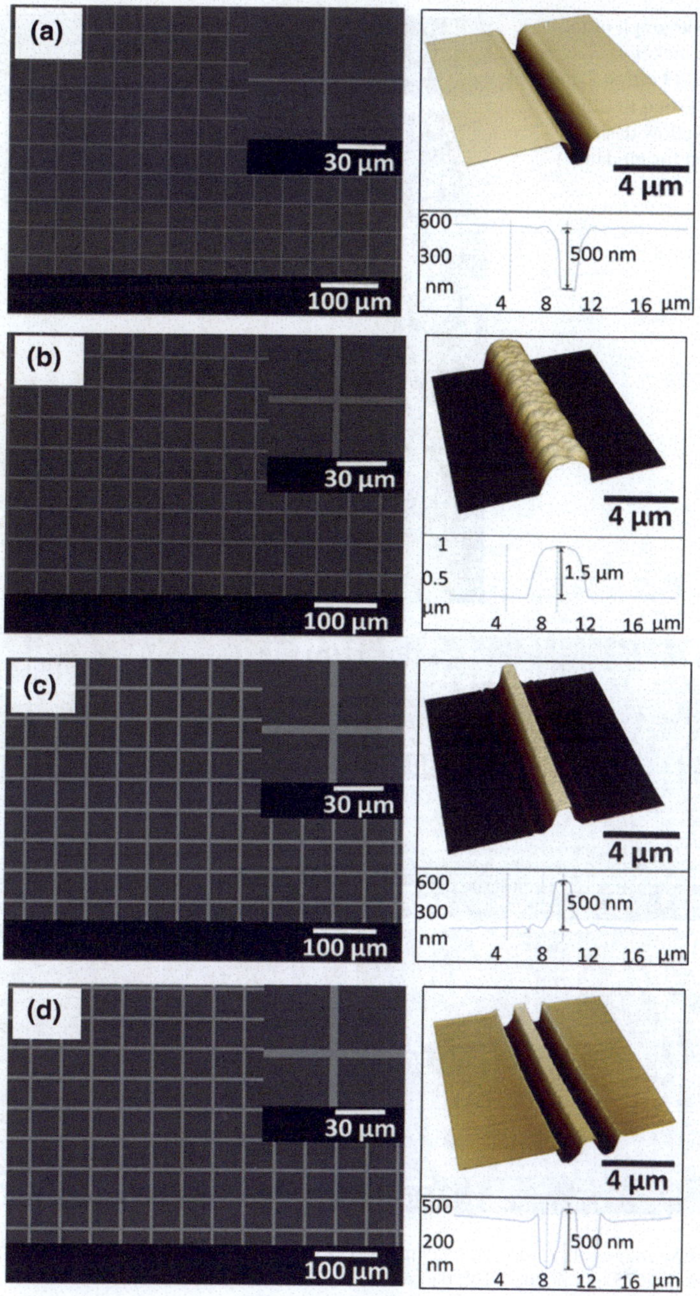

Fig. 4.2 Morphological characterizations by SEM (left) and AFM (right) of prototype 50 μm pitch of nickel micro-EMTE at different fabrication stages of TEIT strategy: **a** As-developed mesh pattern in the reusable SU-8 template, **b** electroplated nickel mesh on the SU-8 template, **c** nickel mesh on the COC film in partial-embedded form, **d** nickel mesh on COC film in fully embedded form

Fig. 4.3 Photograph of the 75 μm pitch nickel micro-EMTE (3 cm × 3 cm) on COC film sited in front of a building (yellow dash outline added for guidance)

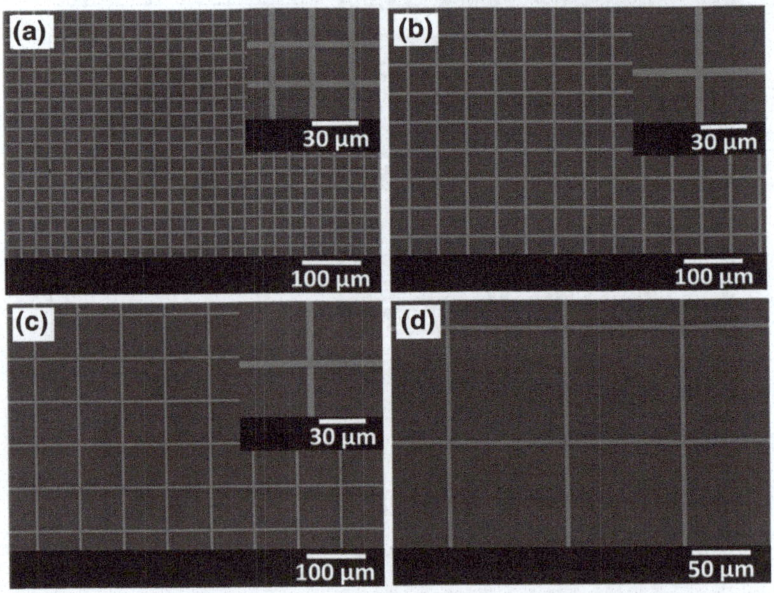

Fig. 4.4 SEM images of prototype nickel micro-EMTEs fabricated by TEIT process **a** 25 μm pitch, **b** 50 μm pitch, **c** 75 μm pitch, **d** 100 μm pitch

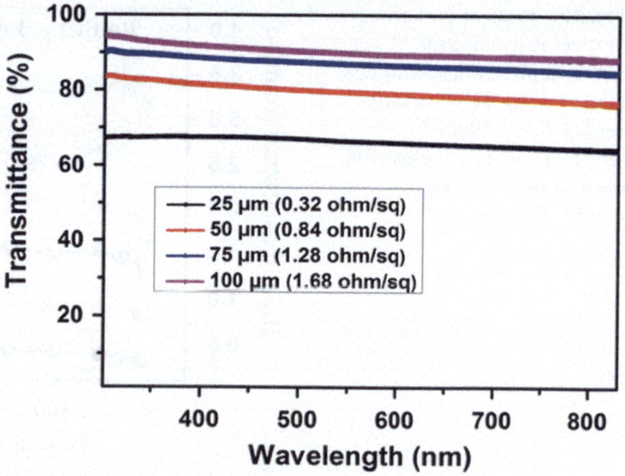

Fig. 4.5 UV–Vis spectra and sheet resistance of the representative nickel micro-EMTEs with pitches of 25, 50, 75, and 100 μm pitch

the versatility of TEIT fabrication. For all these four samples, the trench-width and -depth in SU-8 template were kept constant to be approximately 2 μm and 500 nm respectively. While the metal-mesh had linewidth and thickness of approximately 4 μm and 2 μm, respectively.

The corresponding UV–Vis spectra (in 300–850 nm wavelength range) and sheet resistance of these micro-EMTEs are displayed in Fig. 4.5. It is apparent that the optical transmittance of these samples vary from about 68–93% (at 550 nm wavelength) while sheet resistance changes from 0.32 to 1.68 Ω/sq. Since the metal-mesh thickness for all four sample is constant (approximately 2 μm), therefore these changes in transmittance and sheet resistance are purely due to the variation in pitch of the metal-mesh. Smaller pitch correspond to lower sheet resistance and optical transmittance and vice versa. Thus, it can be concluded from given results that the optical and electrical performance of the micro-EMTEs can easily be tuned in a wide-range just by adjusting the pitch of the metal-mesh. In similar fashion, various combinations of optical transparency and sheet resistance can also be obtained by changing the other geometric parameters of the metal-mesh.

4.3.3 Mechanical Stability of the TEIT Fabricated Micro-EMTEs

From the viewpoint of practical applications, the mechanical flexibility under bending stress is of great importance as discussed in Chap. 3 (Sect. 3.3) as well. Therefore, besides electrical and optical performance measurements on the TEIT fabricated micro-EMTEs, flexibility tests were also performed on all four sample micro-EMTEs (25 μm pitch EMTE, 50 μm pitch EMTE, 75 μm pitch EMTE, and 100 μm pitch EMTE) by repeatedly bending it to a small radius. Figure 4.6 shows the variation in sheet resistance as a function of the number of cycles for repeated compressive

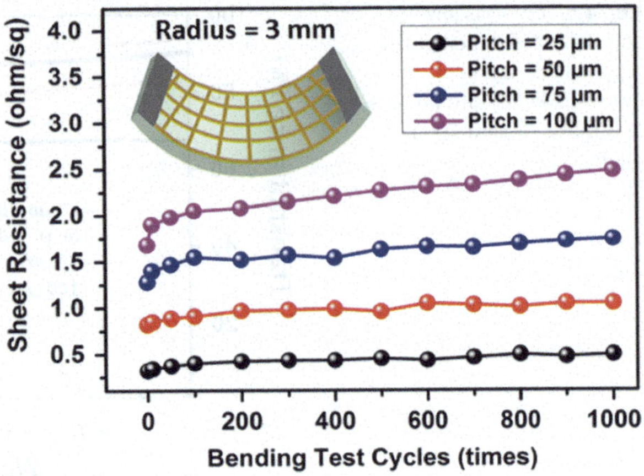

Fig. 4.6 Plots of variations in sheet resistance of different pitch micro-EMTEs fabricated by TEIT strategy versus the number of repeated bending cycles with compressive loading to radius of 3 mm

bending to a radius of 3 mm. The results reveal that for 1000 repeated bendings, the sheet resistances varied from 0.32 to 0.49 Ω/sq for 25 μm pitch EMTE, from 0.82 to 1.05 Ω/sq for 50 μm pitch EMTE, from 1.28 to 1.74 Ω/sq for 75 μm pitch EMTE, and from 1.68 to 2.48 Ω/sq for 100 μm pitch EMTE. The larger increases in sheet resistances of micro-EMTEs with bigger pitches are apparently due to the relatively higher stress on its metal-meshes. However, further studies are required to investigate this trend and the failure mechanisms of the metal-meshes under bending loads. The results further indicate that for all samples, the change in sheet resistances is within 100% of its original value, demonstrating the excellent mechanical stability of micro-EMTEs. This incredible stability of nickel mesh is attributed to its embedded nature in EMTEs.

4.3.4 Reusability of the SU-8 Electrodeposition Templates

SU-8 films are known for its mechanical hardness at elevated temperature and has been used as a mold material for hot-embossing and thermal NIL processes [1–3]. When used as a reusable template in EMTE fabrication, the mechanical and thermal stabilities of SU-8 film are important to maintain its morphology and endure the temperature and pressure effects during thermal imprint transfer step of the TEIT process. We tested the stability of SU-8 template by utilizing it for repeated fabrication cycles in the TEIT process. The SEM and AFM images of the newly fabricated SU-8 template and after its 20 uses are compared in Fig. 4.7a and b respectively. It is clear from these results that there is no noticeable degradation in morphology of SU-8-template after utilizing it for 20 repeated fabrication cycles, showing its potential for reusable electrodeposition template for large-scale manufacturing of micro-EMTEs.

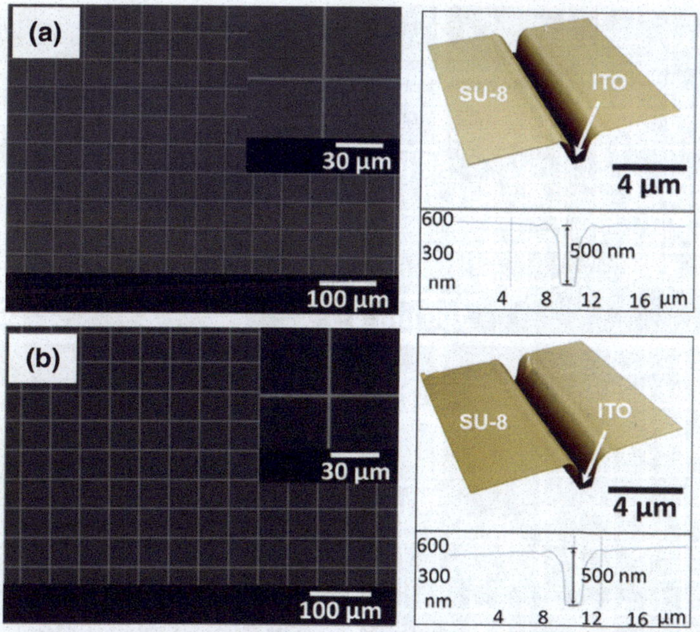

Fig. 4.7 Morphological characterization by SEM (left) and AFM (right) of 50 µm pitch SU-8 template used in TEIT process **a** newly fabricated, **b** after use in 20 production cycles

The structure and performance of the nickel micro-EMTEs fabricated by a single SU-8 template are also analyzed. Figure 4.8a–e presents the SEM images of the 50 µm pitch nickel micro-EMTEs after production cycle 1, cycle 5, cycle 10, cycle 15, and cycle 20. It is apparent from these images that the morphology of these micro-EMTEs is nearly unchanged which demonstrate the excellent reusability of the SU-8 templates.

Similarly, Fig. 4.9 displays the variation in optical transmittance and sheet resistance of nickel micro-EMTEs as a function of the number of times the SU-8 template used in the production cycle of the TEIT process. We can observe that the change in transparency and the sheet resistance of the micro-EMTEs is within 5%, which might arise due to the imprecise control of the experimental parameters during the TEIT process, such as minor variations in electrodeposition area and electrodeposition time etc. This reproducibility of micro-EMTEs by utilizing the same template further confirms the robustness of the SU-8 based electrodeposition templates.

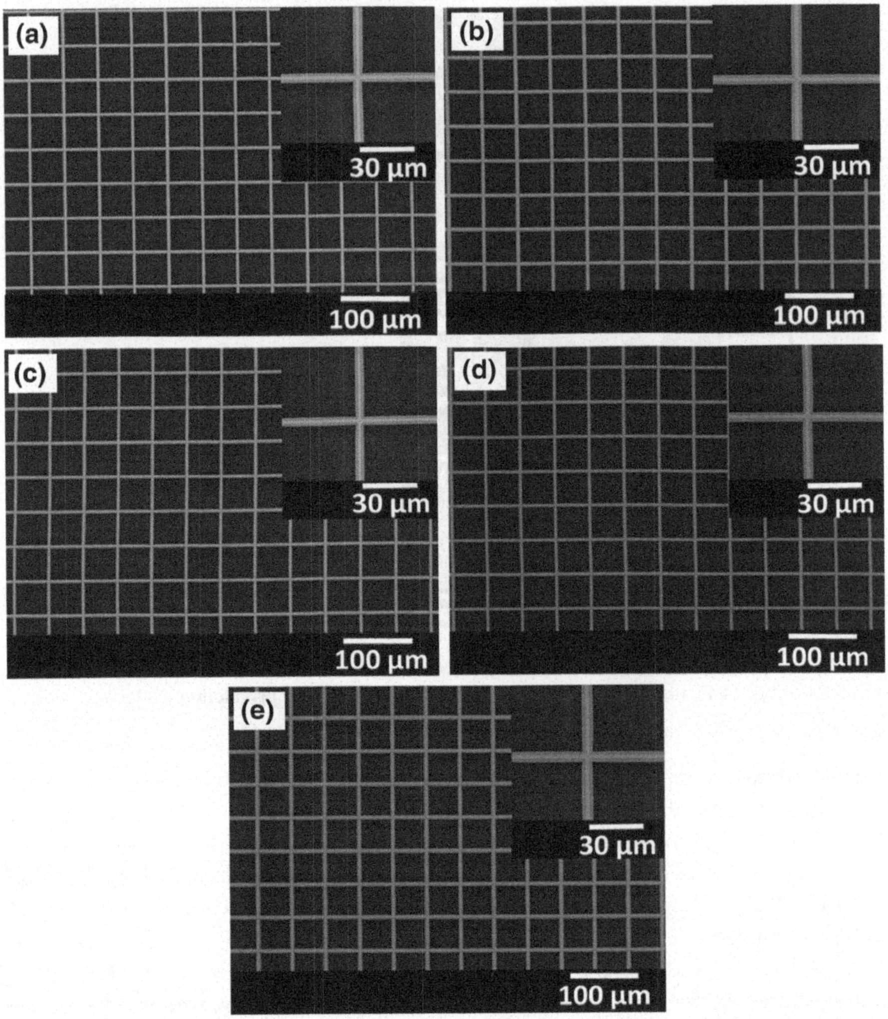

Fig. 4.8 SEM images of nickel micro-EMTEs fabricated by utilizing a single SU-8 template repetitively in TEIT process after unit-production **a** cycle 1, **b** cycle 5, **c** cycle 10, **d** cycle 15, **e** cycle 20

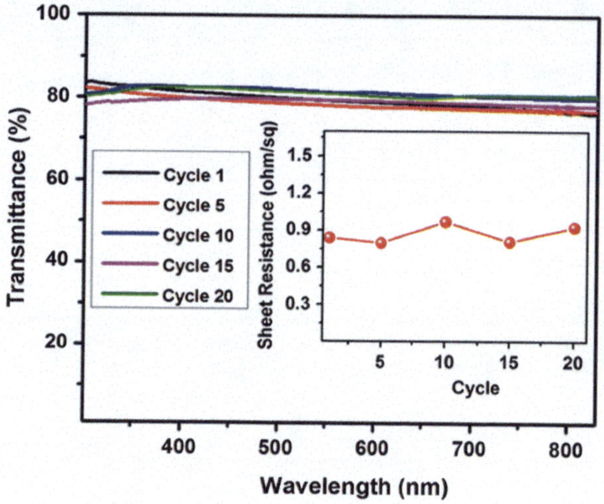

Fig. 4.9 UV–Vis spectra and sheet resistance of the representative nickel micro-EMTEs fabricated by using the same template in TEIT. Inset: plot of variations in sheet resistance versus the number of times the SU-8 template used in the production cycle of the TEIT process

References

1. Bhattacharyya A, Klapperich CM (2006) Thermoplastic microfluidic device for on-chip purification of nucleic acids for disposable diagnostics. Anal Chem 78(3):788–792. https://doi.org/10.1021/ac051449j
2. Liu J, Ro K-W, Nayak R, Knapp DR (2007) Monolithic column plastic microfluidic device for peptide analysis using electrospray from a channel opening on the edge of the device. Int J Mass Spectrom 259(1):65–72. https://doi.org/10.1016/j.ijms.2006.08.017
3. Fan Y, Li T, Lau WM, Yang JA (2012) Rapid hot-embossing prototyping approach using su-8 molds coated with metal and antistick coatings. J Microelectromech Syst 21(4):875–881. https://doi.org/10.1109/JMEMS.2012.2194776

Chapter 5
Nano Embedded Metal-mesh Transparent Electrodes (Nano-EMTEs) Fabricated by LEIT and TEIT Strategies

Abstract This chapter demonstrates the dimensional scalability of LEIT and TEIT fabrication strategies by applying them for making nano-EMTEs. EBL and photolithography with UV stepper are used for nanomesh patterning in LEIT method for making nano-EMTEs. For demonstration, copper nano-EMTEs are fabricated on flexible COC films, exhibiting excellent optical and electrical performances. Similarly in TEIT process, NWL has employed for making the silica template with nano features, which are further used in fabrication of nickel nano-EMTEs on flexible epoxy/COC substrates. The silica templates demonstrated favorable mechanical robustness, and its potential as promising candidate material for reusable electrodeposition templates for production of flexible nano-EMTEs.

5.1 Introduction to Nano-EMTEs

EMTE's structure and its solution-based fabrication strategies (LEIT and TEIT) resolved most limitations of metal-mesh based TCs, as covered in Chaps. 3 and 4 of this dissertation. Despite the advantages offered by the micro-EMTEs, key intrinsic issues of wider linewidths (few microns) and large open spaces (tens of microns) still remained in its structure which cause electrical non-uniformity and is highly undesirable for effective carrier transportation in many organic electronic devices with short exciton diffusion lengths [1–3]. Recently, this issue of poor localized conductivity in metal-mesh based TCs has been addressed by making metal meshes with narrower linewidths (tens of nanometers) and small gaps (few microns) [4–9]. However, the structure and fabrication techniques of these nanomesh TCs still hold the same limitations of expensive vacuum-based metal deposition, non-flat surface topography and weak adhesion between the metal-mesh and flexible substrate. The said limitations demand for new improved fabrication methods for making metal-nanomesh based TCs.

Here we present the dimensional scalability of LEIT and TEIT fabrication strategies by utilizing them for production of nano-EMTEs. In LEIT process for nano-EMTEs, EBL and photolithography with UV stepper are used in nanomesh template patterning step. For demonstration, prototype copper nano-EMTEs are fabricated on

A. Khan, *Novel Embedded Metal-mesh Transparent Electrodes*, Springer Theses,
https://doi.org/10.1007/978-981-15-2918-4_5

flexible COC films, exhibiting excellent optical and electrical performances. Similarly in TEIT process, NWL is utilized for making the reusable electrodeposition templates with nano-features, and are further employed in fabrication of flexible nickel nano-EMTEs.

5.2 Fabrication of Flexible Nano-EMTEs by LEIT Strategy

5.2.1 Experimental Details

In the EBL fabrication, a positive-tone e-beam resist PMMA solution (15k MW, 4 wt% in anisole) was initially spin-coated at 2500 rpm for 60 s onto a cleaned FTO glass substrate and baked on a hotplate at 170 °C for 30 min. The mesh pattern was then written using an SEM (Philips FEI XL30) equipped with a JC Nabity pattern generator. The resist was then developed in a mixed solution of isopropanol and methyl isopropyl ketone at a 3:1 ratio for 1 min. Next, electrodeposition process used commercial aqueous copper solution (Caswell, USA). A Keithley 2400 was used to supply a current density of 3 mA/cm^2 to a two-electrode electrodeposition setup with the PMMA-covered FTO glass as the working electrode and metal bar as the counter electrode. The sample was then rinsed with DI water and immersed in acetone for 5 min to dissolve the PMMA, leaving bare copper mesh on top of FTO glass. Next, the sample was placed on the platen of a hydraulic press (described in Sect. 3.3.1) and covered it with a 100 μm-thick COC (Grade 8007) film. The platens were then heated to 100 °C and 15 MPa of imprint pressure was applied for 5 min. The platens were cooled to 40 °C before releasing the pressure. The COC film was carefully peeled from the FTO glass to obtain the nano-EMTE.

In the UV stepper fabrication, ITO glass substrates (Delta Technologies Ltd. Rs = 10 Ω/sq) were first cleaned and a layer of positive photoresist (Dow Megaposit SPR 220–3.0) was spin-coated at 5000 rpm for 30 s to achieve a thickness of 2 μm. Following a soft baking at 115 °C for 1 min, an i-line stepper (GCA AS200 AutoStep) was used for patterning sub-micron mesh patterns. The photoresist was then developed in AZ MIF-300 developer for 1 min. The samples were then rinsed in water and dried with compressed air. In following electrodeposition and imprint transfer steps, similar protocol as that for micro-EMTEs fabrication was followed (documented in Sect. 3.3.1).

5.2.2 Experimental Results

The EMTE structure and the LEIT fabrication process can be applied to metal meshes with sub-micron feature sizes, which benefit many electronic applications by providing better invisibility and electrical continuity. To demonstrate this dimensional scalability, copper nano-EMTEs were fabricated on COC films by the LEIT process incorporating EBL and a UV stepper. Figure 5.1a–c illustrates the morphological characterization of EBL patterned nano-EMTE fabrication at different stages. Figure 5.1a presents the SEM and AFM images of the trenches created in a PMMA film by EBL. The trench-width and -depth were approximately 400 and 150 nm, respectively. Figure 5.1b displays the copper nanomesh electroplated on the FTO glass (after the removal of the PMMA resist), and Fig. 5.1c shows the imprint transferred copper nanomesh on a COC film. The nanomesh on the COC film was in a fully embedded form, providing stability and strong adhesion with the polymer film.

A UV stepper was also used in the mesh patterning step of LEIT to fabricate copper nano-EMTEs (p = 50 μm, w = 800 nm, t = 1.5 μm) on large area (1.8 cm

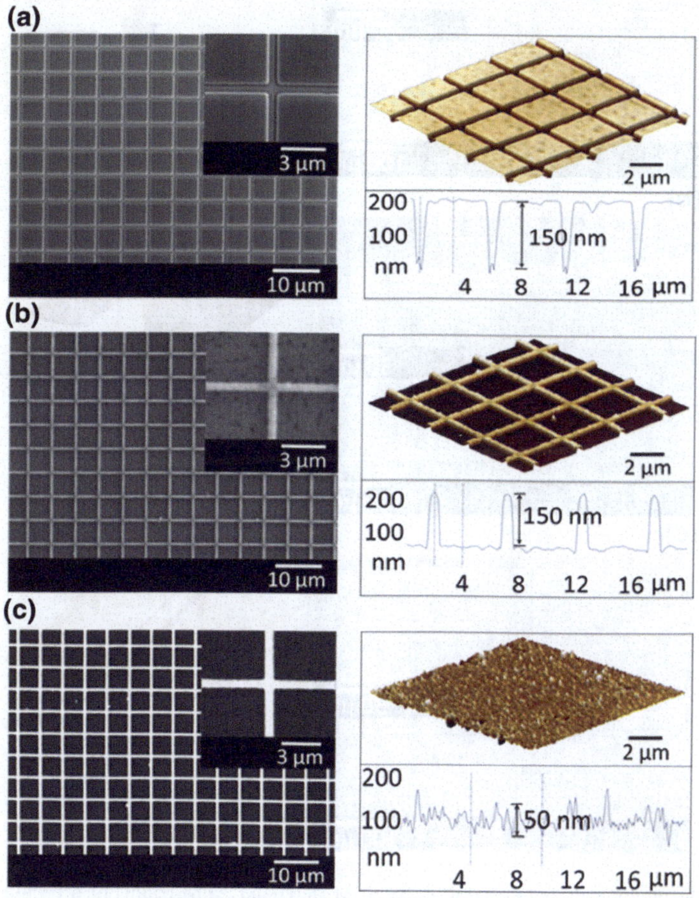

Fig. 5.1 Morphological characterizations by SEM (left) and AFM (right) of a prototype copper nano-EMTE at different fabrication stages. **a** Nanomesh patterns exposed in a PMMA resist film by EBL. **b** Electroplated copper nanomesh on the FTO glass substrate after removal of PMMA resist. **c** Copper nanomesh transferred and fully embedded in a COC film

× 1.8 cm) for electrical and optical characterizations. SEM and AFM characterizations of the nano-EMTE at different stages of the LEIT process are displayed in Fig. 5.2a–c. Figure 5.2a shows the SEM and AFM images of the trench in the photoresist film created using photolithography with UV stepper. Figure 5.2b presents the electroplated copper nanomesh on the ITO glass. As evident from the images, the copper mesh had linewidth and thickness of approximately 900 nm and 1.5 μm,

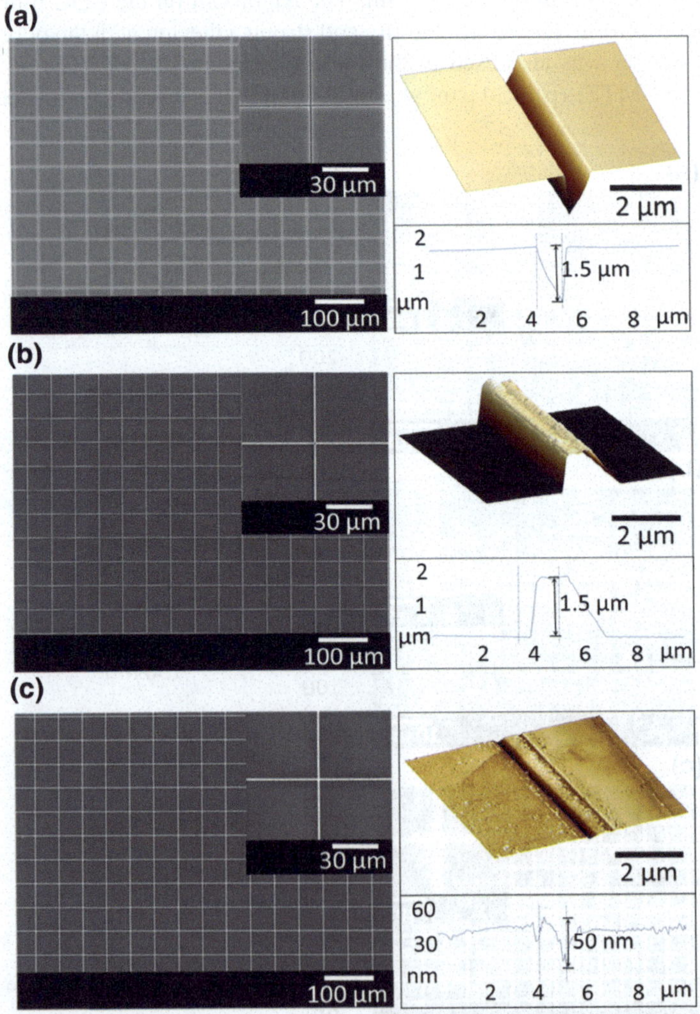

Fig. 5.2 Morphological characterization by SEM (left) and AFM (right) of a prototype copper nano-EMTE at different LEIT fabrication stages. **a** Nanomesh patterns exposed in photoresist film by UV stepper. **b** Electroplated copper mesh on the FTO glass substrate after removal of photoresist. **c** Transferred copper mesh fully embedded in a COC film

Fig. 5.3 UV–Vis spectra and sheet resistance of UV stepper fabricated copper nano-EMTE (50 μm pitch and 900 nm linewidth) on COC film

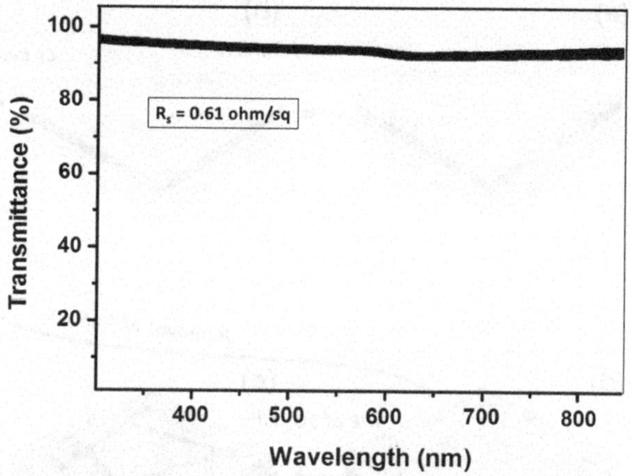

respectively. The AFM characterization (shown in Fig. 5.2c) confirms the embedded structure of nano-EMTE by revealing that its final roughness is less than 50 nm.

Figure 5.3 presents the UV–Vis spectra and sheet resistance of this UV stepper fabricated nano-EMTE. Due to the smaller linewidth, the sample demonstrates high optical transmittance (94% at 550 nm wavelength) and very low sheet resistance (0.61 Ω/sq). These successful fabrications confirmed that the EMTE structure and key fabrication steps in LEIT can be reliably scaled down to a sub-micron range. In fact, the electrodeposition and imprint-transfer processes allow an ultimate linewidth of metal-mesh in the sub 100 nm range [10] with a high aspect ratio, providing great potential of further reducing the linewidth of the EMTE mesh for better electrical continuity and transmittance.

5.3 Fabrication of Flexible Nano-EMTEs by TEIT Strategy

As discussed in Chap. 4, the improved TEIT technique simplify the fabrication of micro-EMTEs by removing the mandatory patterning step, yet key issues of wider linewidths and large open spaces still remained in its structure which also call for scale-down TEIT approach for making nano-EMTEs. This section presents the details of TEIT fabrication of nano-EMTEs.

5.3.1 Experimental Details

The TEIT fabrication requires a reusable electrodeposition template. Therefore, first, a silica-template with nano features was fabricated for electrodeposition step in the TEIT fabrication. Figure 5.4 schematically explains the methodology of silica-

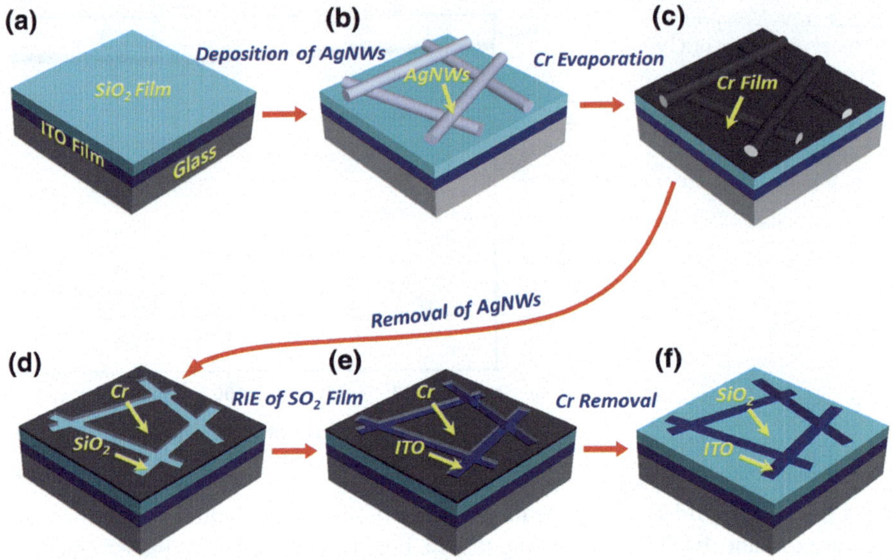

Fig. 5.4 Schematic illustration of silica-template fabrication procedure. **a** Deposition of SiO_2 film on ITO glass substrate. **b** Distribution of AgNWs on SiO_2 film. **c** Evaporation of Cr film. **d** Removal of AgNWs. **e** Reactive ion etching of SiO_2 film. **f** Removal of the Cr film marks the completion of reusable electrodeposition template with nano features

template fabrication. ITO glass substrate (3 cm $\times$ 3 cm) was cleaned, following the same protocol as described in Chaps. 3 and 4 of this dissertation. Next, 80 nm thick SiO_2 film was deposited by radio frequency (RF) sputtering using a 2 inch diameter SiO_2 target in argon atmosphere in the pressure range of 5 mTorr and RF power of 150 W (Fig. 5.4a). Then, AgNWs were spread on top of the SiO_2 film (Fig. 5.4b) using water-bath assisted convective assembly technique [11]. The employed AgNWs (purchased from ACS Material, USA) have an average length and diameter of nearly 20 μm and 125 nm, respectively. During the process, a homebuilt dip-coater with optimized pull-out speed of 2 mm s^{-1} was used to transfer the assembled AgNWs to the sample. Subsequently, chromium (Cr) film of 30 nm thickness was deposited onto the sample by thermal evaporation (Fig. 5.4c). The distance between the sample and the Cr source was about 30 cm and the deposition rate was 5 Å s^{-1}. Next, the AgNWs were removed by embedding it in a poly-di-methyl siloxane (PDMS) film. Sylgard 184 silicone elastomer base and a curing agent (Dow Corning, USA) were mixed in a ratio of 10:1 by mass to form PDMS pre-polymer mixture. The pre-polymer was poured onto the sample and degassed at 5–10 Pa in a desiccator to remove any air bubbles. The PDMS pre-polymer was then cured for 1 h at 100 °C on a hot plate. After curing, the PDMS film was peeled off from the sample, taking along all the AgNWs and leaving its traces as openings in the Cr film (Fig. 5.4d). Then, the SiO_2 film was selectively dry-etched by O_2/CHF_3 (5sccm:20sccm) gas using Cr film as a mask. Dry-etching was performed with RF power of 200 W using a home-built reactive ion etching (RIE) system containing a plasma generator (TONSON, Taiwan) with homemade gas controller. The base pressure of the chamber was 20 Pa. The etch rate of SiO_2 was optimized to be 15 nm min^{-1} using the above conditions. After etching of SiO_2 and selectively exposing the ITO film (Fig. 5.4e), the Cr film was removed by immersing the sample in Cr etchant solution (Techni Etch Cr01, Technic, France) for 10 min at room temperature (Fig. 5.4f). Finally the template was rinsed in DI water and blow-dried with compressed air.

Prototype nickel nano-EMTEs are fabricated by TEIT process using the silica-template (Fig. 5.4f). The fabrication process is schematically demonstrated in Fig. 5.5. In a typical fabrication repetition, metal was grown in the trenches of the template by electrodeposition (Fig. 5.5b). Electrodeposition was performed using commercial nickel plating solution (Caswell, USA). Keithley 2400 SourceMeter was utilized to provide a current density of 3 mA/cm^2 to a two-electrode electrodeposition system with the silica-template as the working electrode and metal bar as the counter electrode. Electrodeposition time was optimized to be 3 min to reach the average overplated metal thickness of 600 nm, required for subsequent successful transfer. After the metal electrodeposition, the sample was thoroughly rinsed in DI water and blow-dried with compressed air. Next, the metal-mesh was transferred to a flexible substrate by UV imprint process (Fig. 5.5c). During the process, NOA-61 UV-curable epoxy (Norland Products Inc., USA) was drop-casted on the electroplated silica-template and then a 100 μm thick COC (Grade 8007) film was placed on top of it. The epoxy/COC substrate was exposed to a UV light (lamp power =

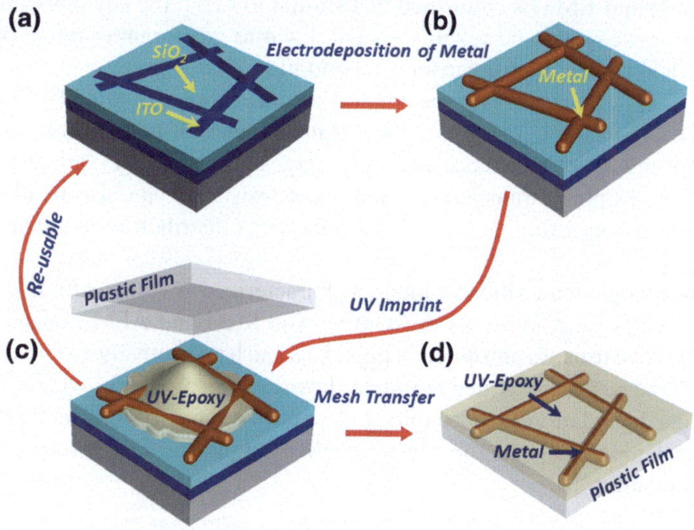

Fig. 5.5 Schematic illustration of TEIT fabrication strategy for nano-EMTEs. **a** Silica-template with nano features on ITO glass substrate. **b** Electrodeposition of metal inside the silica trenches to form a uniform metal mesh. **c** Pressing and exposing the stack of epoxy/COC substrate and electroplated silica-template to UV light. **d** Peeling off the epoxy/COC substrate with the metal mesh transferred in embedded form

400 W, time = 3 min, peak wavelength = 385 nm) with an applied pressure of approximately 5 bars to press the silica-template and the epoxy/COC substrate. The template/substrate stack was then removed from the imprint press after releasing the pressure. Finally, the epoxy/COC substrate was manually peeled off from the silica-template, taking along the metal nano-mesh in embedded form (Fig. 5.5d), and the template was ready for the next production cycle.

5.3.2 *Experimental Results*

A nano-EMTE with a nickel mesh on epoxy/COC substrate was fabricated by TEIT strategy as a demonstration. Figure 5.6 displays the morphological characterization of this nano-EMTE at different stages of the fabrication. Figure 5.6a shows the SEM and AFM images of the trench in the silica-template created using NWL (Fig. 5.5a). On this sample, the trench-width and -depth are approximately 150 and 80 nm respectively, and its average pitch over whole area of the template is less than 10 μm. Figure 5.6b demonstrates the electroplated nickel mesh on the silica-template (Fig. 5.5b). As apparent from the results, the nickel mesh has linewidth and thickness of nearly 900 nm and 700 nm, respectively. Figure 5.6c reveals the transferred nickel mesh onto epoxy/COC substrate (Fig. 5.5d). The AFM characterization displays that the surface roughness of the completed nano-EMTE is 80 nm which is approximately equal to the depth of the trench in the silica template. This successful fabrication of nano-EMTEs confirmed that similar to LEIT, the key fabrication steps in TEIT process can also be reliably scaled down to a nanometer range providing better electrical continuity for carrier transportation.

Figure 5.7 shows the UV–Vis spectra and sheet resistance of this TEIT fabricated nano-EMTE. The sample exhibits optical transmittance of nearly 80% (at 550 nm wavelength) and sheet resistance of 1.84 Ω/sq. similar to LEIT process, various combinations of optical transparency and sheet resistance can also be obtained in TEIT process by adjusting the density of nanowires distribution in silica template fabrication.

The morphology of the silica template is characterized after using for five repeated fabrication cycles to examine its reusability. The SEM and AFM images of silica template after five uses are presented in Figs. 5.8a and b respectively. Comparing these images with the newly fabricated template shown in Fig. 5.6a, it is evident that there is no obvious degradation in morphology after utilizing for five repeated fabrication cycles. This demonstrates the robustness of silica template in TEIT process for nano-EMTEs fabrication.

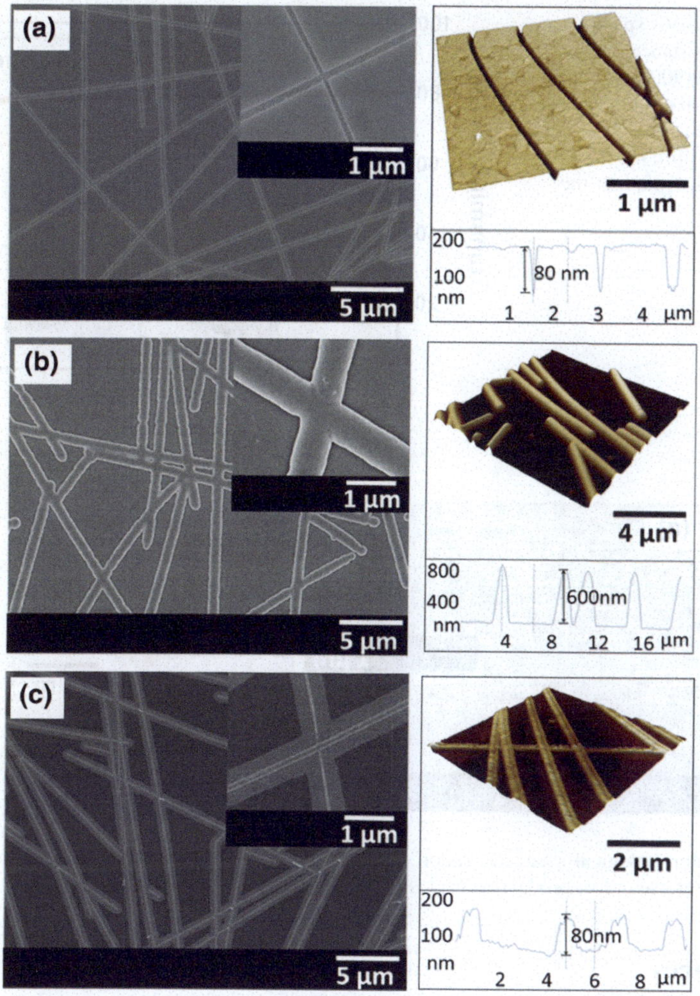

Fig. 5.6 Morphological characterization by SEM (left) and AFM (right) of prototype nickel nano-EMTEs at different fabrication stages of TEIT process: **a** Mesh pattern in the silica-template. **b** Electroplated nickel mesh on the silica-template. **c** Nickel nanomesh transferred and embedded in the epoxy/COC substrate

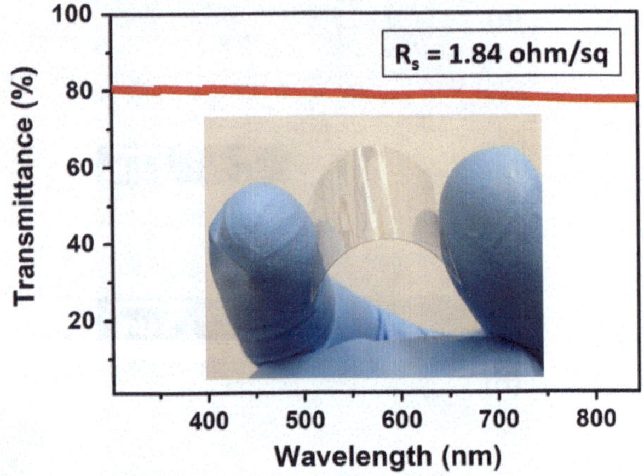

Fig. 5.7 UV–Vis spectra and sheet resistance of nickel nano-EMTE (900 nm linewidth, 700 nm thickness) on 3 × 3 cm large epoxy/COC film; the inset show the photograph of the final EMTE

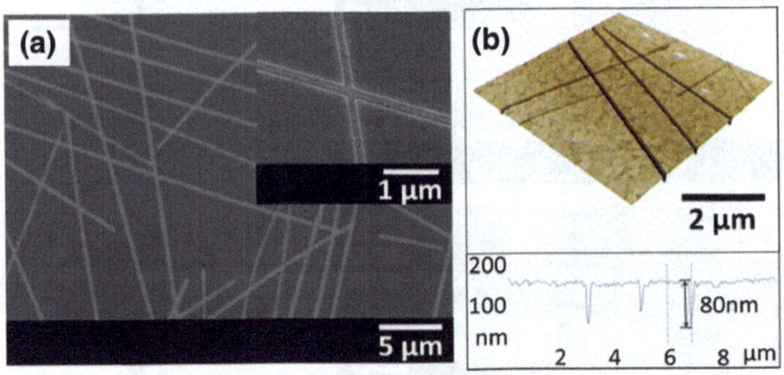

Fig. 5.8 Morphological characterization [SEM (**a**) and AFM (**b**)] of silica electrodeposition template after use in five production cycles of TEIT process

References

1. Deng B, Hsu P-C, Chen G, Chandrashekar BN, Liao L, Ayitimuda Z, Wu J, Guo Y, Lin L, Zhou Y, Aisijiang M, Xie Q, Cui Y, Liu Z, Peng H (2015) Roll-to-roll encapsulation of metal nanowires between graphene and plastic substrate for high-performance flexible transparent electrodes. Nano Lett 15(6):4206–4213. https://doi.org/10.1021/acs.nanolett.5b01531
2. Gao T, Huang P-S, Lee J-K, Leu PW (2015) Hierarchical metal nanomesh/microgrid structures for high performance transparent electrodes. RSC Adv 5(87):70713–70717. https://doi.org/10.1039/C5RA14851K
3. Gao T, Li Z, Huang P-S, Shenoy GJ, Parobek D, Tan S, Lee J-K, Liu H, Leu PW (2015) Hierarchical graphene/metal grid structures for stable, flexible transparent conductors. ACS Nano 9(5):5440–5446. https://doi.org/10.1021/acsnano.5b01243
4. Hak-Jong C, Soyoung C, Pil-Hoon J, Ju-Hyeon S, Yang-Doo K, Heon L (2015) Uniformly embedded silver nanomesh as highly bendable transparent conducting electrode. Nanotechnology 26(5):055305
5. Gong S, Zhao Y, Yap LW, Shi Q, Wang Y, Bay JAPB, Lai DTH, Uddin H, Cheng W (2016) Fabrication of highly transparent and flexible nanomesh electrode via self-assembly of ultrathin gold

nanowires. Adv Electron Mater 2(7):1600121-n/a. https://doi.org/10.1002/aelm.201600121
6. Jang S, Jung W-B, Kim C, Won P, Lee S-G, Cho KM, Jin ML, An CJ, Jeon H-J, Ko SH, Kim T-S, Jung H-T (2016) A three-dimensional metal grid mesh as a practical alternative to ITO. Nanoscale 8(29):14257–14263. https://doi.org/10.1039/C6NR03060B
7. Qiu T, Luo B, Ali F, Jaatinen E, Wang L, Wang H (2016) Metallic nanomesh with disordered dual-size apertures as wide-viewing-angle transparent conductive electrode. ACS Appl Mater Interfaces 8(35):22768–22773. https://doi.org/10.1021/acsami.6b08173
8. Jo HS, An S, Lee J-G, Park HG, Al-Deyab SS, Yarin AL, Yoon SS (2017) Highly flexible, stretchable, patternable, transparent copper fiber heater on a complex 3D surface. NPG AsiaMater 9:e347. https://doi.org/10.1038/am.2016.206
9. Yu S, Han HJ, Kim JM, Yim S, Sim DM, Lim H, Lee JH, Park WI, Park JH, Kim KH, Jung YS (2017) area-selective lift-off mechanism based on dual-triggered interfacial adhesion switching: highly facile fabrication of flexible nanomesh electrode. ACS Nano 11(4):3506–3516. https://doi.org/10.1021/acsnano.7b00229
10. Burek MJ, Greer JR (2010) Fabrication and microstructure control of nanoscale mechanical testing specimens via electron beam lithography and electroplating. Nano Lett 10(1):69–76. https://doi.org/10.1021/nl902872w
11. Duan S-K, Niu Q-L, Wei J-F, He J-B, Yin Y-A, Zhang Y (2015) Water-bath assisted convective assembly of aligned silver nanowire films for transparent electrodes. Phys Chem 17(12):8106–8112. https://doi.org/10.1039/C4CP05989A

Chapter 6
Applications of Embedded Metal-mesh Transparent Electrodes in Flexible Electronic Devices

Abstract This chapter summarizes the applications of EMTEs in flexible bifacial DSSCs and FTTHs. A highly efficient nanostructured flexile transparent CE fabricated by a facile and low-cost method for DSSC devices is reported. This hybrid CE consists of a nickel micro-EMTE with in situ decorated PtNPs and fully embedded in a flexible transparent substrate, and demonstrates superior optoelectronic performances and excellent mechanical flexibility under bending stress. The in situ loading of PtNPs avoids the wastage of expensive platinum material, minimizes loss in optical transparency and facilities the high electro-catalytic activity for triiodide reduction. The synergetic features of the proposed hybrid PtNP-EMTE used as CEs for flexible DSSCs enable improved power conversion efficiency. When a Ti-foil is used as the photo-anode, unifacial flexible DSSCs with PtNP-EMTE CE exhibits an excellent PCE of 6.91%. In flexible bifacial DSSCs using PtNP-coated ITO-PEN as photo-anode and the PtNP-EMTE as CE, remarkable PCEs of 5.67% (front-side illumination) and 4.87% (rear side illumination) are recorded. The front-to-rear PCE ratio of our bifacial DSSC approaches 85%, which is among the highest in published literature. These promising results show the great potential of this hybrid transparent CE in scalable production and commercialization of low-cost and efficient flexible DSSCs. Based on LEIT fabricated micro-EMTEs, a transparent and flexible thin-film heater is also fabricated, demonstrating excellent performance compared with existing products. With superior performances, our micro-EMTEs and nano-EMTEs are potential candidates for other flexible electronic devices as well.

6.1 Introduction

Flexible electronic devices, such as solar cells, [1] light-emitting diodes, [2] displays, [3] touchscreens, [4] smart windows, [5] energy harvesters, [6] batteries, [7] supercapacitors, [8] and transparent heaters [9] etc. are generally built on substrates like plastics or metallic foils that can be bended, folded, twisted, and rolled with minor effects on its electronic performance. Flexible electronic devices are usually made using layer by layer fabrication approach, where thin layers of active electronic materials are deposited on top of the flexible TCs. In many applications, extra

A. Khan, *Novel Embedded Metal-mesh Transparent Electrodes*, Springer Theses,
https://doi.org/10.1007/978-981-15-2918-4_6

layers are also utilized for encapsulation and packaging. For better performance, TCs employed in these flexible electronic devices require three properties simultaneously, which are high electrical conductivity, high optical transparency, and high mechanical flexibility. To achieve these characteristics, new-generation TCs based on organic and inorganic materials are extensively investigated. Likewise, we also utilized our EMTEs in flexible electronic devices as practical applications.

This chapter presents the applications of EMTEs in flexible bifacial DSSCs and flexible transparent heaters. Based on our micro-EMTE, a novel CE is developed for flexible bifacial DSSCs. This novel CE features a micro-EMTE with catalytic PtNPs in situ electrodeposited only on the surface of the nickel mesh without considerably reducing its optical transparency. This hybrid PtNP-decorated nickel EMTE demonstrates one of the best bifacial flexible DSSC in terms of PCE. Also, a flexible and transparent thin-film heater based on the micro-EMTEs is fabricated and characterized showing a rapid response, requires low input power density, and operates at an ultra-low voltage, clearly demonstrating the potential of our EMTEs as high-performance flexible transparent heaters.

6.2 Micro-EMTEs Based Flexible Bifacial DSSCs

Flexible solar cells have attracted substantial attention over the last decade because they have shown great promise for a variety of applications due to their unique characteristics such as low weight, bendability, and portability. As compared to other types of solar cells, flexible DSSCs fabricated on conductive plastic substrates are considered particularly promising for their simple structure, low cost, high manufacturing throughput, and suitability as portable and mobile ubiquitous power sources [10]. In recent years, considerable studies have been carried out to improve the PCE of flexible DSSCs. However, due to practical challenges brought by intrinsic properties of the flexible substrates, such as low processing temperature, PCE of flexible DSSCs is still much lower than that of rigid DSSCs, [11–13] resulting in increased energy production cost and payback time. Therefore, further reduction of cost and improvement of PCE by employing new materials and new device structures in flexible DSSCs are remaining concerns for their large-scale commercial applications [14]. One solution to this could be the design of bifacial DSSCs, which may almost double the light harvesting capability [15–18] by utilizing the incident light from both sides. Especially, when applied with a reflecting background such as a static concentrator, the energy output efficiency can be significantly enhanced [15, 19]. Besides more efficient energy harvesting, this kind of devices also enables unique applications such as in electricity-generating windows [20]. However, unlike conventional DSSCs, in such device architecture an efficient transparent CE is obligatory to utilize the rear-side illumination, which makes its fabrication difficult and expensive [21]. The challenges get further intensified when flexibility also comes into the demands. Therefore, the development of low-cost flexible transparent CEs is the key challenge for the commercial realization of bifacial flexible DSSCs [21–23].

Currently, the dominant material for flexible CEs is TCO coated plastic film, particularly ITO or FTO coated PEN or PET substrates, with the surface covered by PtNPs [24–26] for their superior electro-catalytic performance in reducing I_3^- to I^-. [27, 28] Although recent efforts have been made on developing alternative Pt-free CEs in DSSCs, [29–38] their conductivity and catalytic activity are still inferior and not yet ready for practical applications [26, 28]. However, even though PtNP-based plastic CE has exhibited desirable catalytic performance, many concerns that mainly arise from the TCO-coated plastic film, including material cost, low abundance, film fragility, thermal stability under high-temperature processing and weak adhesion of the PtNPs, need to be addressed in order to make it suitable for use in large-scale commercial applications of plastic DSSCs [14, 39]. In addition, deposition of PtNPs on the CE substrates also need to be examined and improved. Sputtering [40, 41] of Pt has been the preferred choice for platinizing plastic conductive substrates, while several other techniques, such as atomic layer deposition, [42] dipping, [25] spray coating, [43] electrochemical deposition, [44] and chemical reduction [39] were recently developed as alternatives to sputtering in order to overawe some of its drawbacks relating to the high cost and low throughput due to the required high vacuum processing [45]. However, most of these novel techniques still cover the whole substrate area with PtNPs, leading to inevitable wastage of expensive Pt material and consequently increasing the cost and the energy payback time, which are limiting factors for DSSC's large-scale commercial applications. Furthermore, PtNPs deposited on the full substrate surface are typically opaque and significantly reduce the optical transmittance of the transparent conductive substrate, resulting in poor efficiency under rear illumination through the CE side. These limitations in existing PtNP-based CEs demand for improved and low-cost fabrication methods for new CEs with better performance.

TCs based on metal-meshes [46–49] have shown superior transparency, conductivity, and mechanical flexibility. They have been increasingly used in a variety of solar cells [50–55] and even in large-scale solar modules [56, 57]. Their capability of reaching very low sheet resistance in the 0.1–1 Ω/sq range is particularly important for large-size solar cell modules requiring reduced voltage drop across the whole CE. Therefore, it is potentially advantageous to combine the excellent electrical, optical and mechanical traits of the metal-mesh based TCs with the high electro-catalytic activity of PtNPs, to develop an ideal alternative flexible transparent CEs for bifacial flexible DSSCs [58].

Based on our micro-EMTE [49, 59, 60], here we developed a novel CE for flexible bifacial DSSCs. This novel CE features a thick nickel micro-mesh fully embedded and anchored in a highly transparent flexible COC substrate, with catalytic PtNPs in situ electrodeposited only on the surface of the nickel mesh without considerably reducing its optical transparency. This composite electrode is fabricated through a scalable, facile, and fully solution-processed procedure, and shows enhanced opto-electronic performances and superior mechanical flexibility under bending stresses, all among the best results published so far. Cyclic voltammogram (CV) and electrochemical impedance spectroscopy (EIS) are used to evaluate its electrochemical performance as CE in DSSCs. This hybrid PtNP-decorated nickel micro-EMTE demonstrates excellent PCE of 6.91% when used as the CE in a flexible DSSC with a Ti-foil photoanode, exhibiting superior performance comparing to a control device with a traditional ITO-PEN CE. When used in a flexible bifacial DSSC with a transparent ITO-PEN photoanode, this hybrid PtNP-coated nickel micro-EMTE achieves power conversion efficiencies of 4.87% under rear-illumination and 5.67% under front-illumination, with an 85% ratio between them, all among the best results reported for bifacial flexible DSSCs.

6.2.1 Experimental Details

6.2.1.1 Fabrication and Characterization of Hybrid PtNP-EMTE

First, nickel micro-EMTEs are prepared by LEIT and TEIT fabrication approaches as explained in Chaps. 3 and 4. The PtNPs were subsequently deposited on nickel micro-EMTEs using commercial aqueous platinum plating solution (Metalor, Met-Pt 200S, Switzerland) using pulse electro-deposition method by applying the pulse waveform between the onset voltage and a higher overpotential via electrochemical workstation (CHI 660E, USA) with three-electrode system.

The morphologies of samples were characterized by a SEM (Hitachi S-4800, Japan) and AFM (Bruker MultiMode-8, USA). The sheet resistances of the samples were measured by the four-probe method to eliminate contact resistance. During the measurement, four electrodes are placed at four corners of a square (3 cm × 3 cm) of the sample and the resistance is recorded with a Sourcemeter (Keithley 2400, USA). Optical transmission spectra was measured by a UV-Vis spectrometer (PerkinElmer Lambda 25, USA). For electrochemical durability evaluation, CV was conducted using an electrochemical workstation (CHI 660E, USA) with three-electrode system utilizing 0.5 cm^2 CE, Pt wire (F¼ 0.3 mm) and Pt plate (1 cm^2) served as working electrode, reference electrode and counter electrode, respectively. The electrolyte for CV test contained 50 mM LiI (Lithium iodide, Acros), 10 mM I_2 (Iodine, Sigma), 500 mM $LiClO_4$ (Lithium perchlorate, Acros) in 3-MPN (3-methoxypropionitrile, Sigma), and the scan rate for measurement was 5 mV/s. The catalytic activity toward tri-iodide reduction was measured by electrochemical impedance spectrometer (Autolab PGSTAT320N, Netherlands) with alternating current voltage ranging from 10^{-1} to 10^6 Hz with 10 mV amplitude in a symmetric cell.

6.2.1.2 Fabrication and Characterization of Flexible DSSCs

The ITO-PEN (200 μm thick, ≈13 Ω/sq, transmittance 80%) was first cleaned using acetone under sonication for 10 min and then treated by UV-O_3 to increase the wettability and improve the adhesion between low-temperature TiO_2 paste and the substrate. The mesoporous and binder-free TiO_2 layer containing particles with sizes ranging from 50 to 400 nm was then coated on ITO-PEN by doctor-blade method to reach TiO_2 thickness of 7 μm. The wet film was dried in an oven at 70 °C to improve particles necking. In order to further eliminate water, the sample was subsequently heated at 120–150 °C. For the Ti-TiO_2 based flexible DSSCs, the titanium substrate was polished and pretreated with 90 °C H_2O_2 solution for 20 min. It was subsequently screen-printed with TiO_2 paste (particle size ≈20 nm, Eternal, Taiwan) until the film thickness reached 10 μm, and then sintered in a furnace at 450 °C for 30 min to produce the nanoporous TiO_2. The TiO_2-coated film was sensitized by Ru complex dye (Dyesol N719, Australia) in ethanol at room temperature in a 0.4 mM dye solution for 4 h (for ITO-PEN anode)/12 h (for Ti-anode) to complete the dye impregnation. The effective area of the TiO_2 photoanode is 0.23 cm^2. The dye-adsorbed photoanode and the hybrid PtNP-EMTE were stacked face-to-face and sealed with a 30-μm-thick thermal-plastic Surlyn spacer (SX1170-25, Solaronix, Switzerland). For DSSCs with ITO-PEN photoanode, a proper amount of liquid electrolyte that contains 0.4 M TBAI (tetrabutylammonium iodide, Sigma), 0.3 M NMB (N-methylbenzimidazole, Sigma), 0.4 M LiI and 0.04 M I_2 in mixture of AN and 3-MPN (1:1, v/v) was injected into the gap between the two electrodes. For DSSCs with Ti-foil photoanode, electrolyte containing 0.6 M PMII (1-methyl-3-propylimidazolium iodide, Sigma), 0.05 M I_2, 0.1 M LiI, 0.5 M TBP (tributyl phosphate, Sigma) in AN/VN (85:15 = v/v) was injected. The J-V curves of DSSCs were recorded with a computer-controlled digital source meter (Keithley 2400, USA) under a standard solar simulator (Peccell, PEC-L01, Japan) for 1 sun illumination (AM 1.5G, 100 mW cm^{-2}).

6.2.2 Experimental Results

6.2.2.1 Structure of Bifacial Flexible DSSC with PtNP-EMTE Counter Electrode

The flexible bifacial DSSCs developed in this work feature a commonly used architecture that consists of an ITO-PEN photoanode, dye-soaked TiO_2 layer, electrolyte, and a CE, from the front-side to the rear-side. The CE is the novel PtNP-decorated nickel EMTE. The structure and the assembly process of the device are schematically described in Fig. 6.1. The nickel EMTE shown in Fig. 6.1a is prepared by both LEIT and TEIT approaches. The hybrid PtNP-EMTE in Fig. 6.1b is obtained by subsequently depositing PtNPs using pulse electrodeposition method by applying the pulse waveform between the onset voltage and a higher over-potential (Fig. 6.12, Appendix). The photoanode is fabricated by coating mesoporous and binder-free TiO_2 colloidal solution on a flexible ITO-PEN substrate (Fig. 6.1c) by the doctor-blade technique. After drying, the film is dipped and sensitized in the dye solution at room temperature, and turned into deep brown color (Fig. 6.1d). As shown in Fig. 6.1e, the dye-adsorbed photoanode and the hybrid PtNP-EMTE are then assembled and sealed to form a cavity before liquid electrolyte is filled into complete the device fabrication.

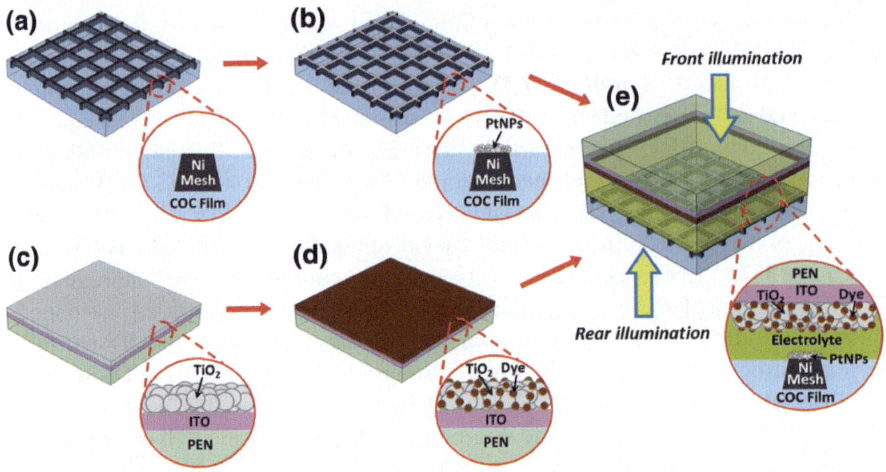

Fig. 6.1 Schematic illustration of the fabrication process. **a** Nickel EMTE on COC film; **b** deposition of PtNPs on nickle micro-EMTE by puls electro-deposition; **c** mesoporous TiO_2 layer on ITO-PEN coated by doctor-blade technique; **d** sensitizing TiO_2 film by dipping in dye solution; and **e** final structure of the assembled device containing the dye-adsorbed photo-anode, liquid electrolyte and the PtNP-EMTE

6.2.2.2 Electrical, Optical and Mechanical Characterization of PtNP-EMTE

The nickel micro-EMTEs used in this study is prepared by transferring a 50-μm-pitch nickel mesh with 1.2 μm thickness and 3 μm linewidth (Fig. 6.13, Appendix) to a highly transparent flexible COC substrate (Fig. 6.14, Appendix). Figure 6.2a presents the EDS mapping and AFM imaging of the fabricated nickel micro-EMTE. It is evident from these images that the mesh, which is of pure nickel, is fully embedded in the COC substrate. The nickel EMTE shows an excellent sheet resistance of $R_{Sh} = 1.32\ \Omega/sq$ along with an optical transparency of 74% in the wavelength range between 300 and 800 nm, which covers the absorption window of most dye solutions used in DSSCs. Typically, our micro-EMTEs have shown very high FoM, therefore EMTEs serve as excellent platform for developing novel CEs for DSSCs. The hybrid PtNP-EMTE was then obtained by depositing PtNPs onto the nickel micro-EMTE using pulsed electro-deposition technique. Pulse width and potential have been optimized to uniformly deposit PtNPs of consistent sizes (Fig. 6.15, Appendix) only on the exposed surface of the thick embedded nickel mesh. Figure 6.2b displays the EDS mapping and AFM image of the nickel mesh coated with a thin 100 nm thick film of PtNPs, confirming its purity and uniform distribution. The details of the EDS analysis are provided in Fig. 6.16 (Appendix).

Since the PCE of the devices under rear illumination is directly related to the amount of incident light, high optical transmittance of the PtNP-coated CE is important to achieve satisfactory PCE. Comparing to a traditional transparent CE using ITO-PEN substrates covered by PtNPs, our PtNP-EMTE is advantageous because

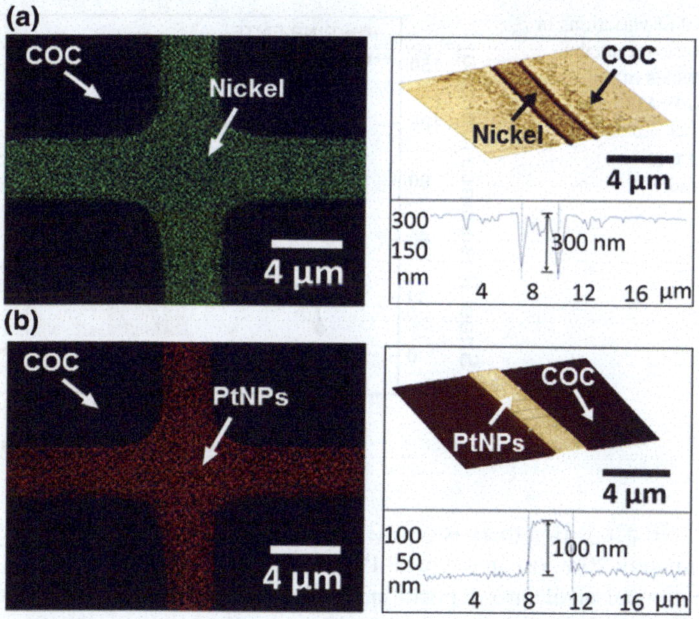

Fig. 6.2 Material and morphological characterizations by EDS-SEM (left) and AFM (right) of the nickel micro-EMTE on COC film. **a** Before PtNPs deposition. **b** After PtNPs deposition

the opaque PtNPs are only deposited on the exposed surface of nickel mesh therefore will not significantly sacrifice the optical transmittance of the CE. Figure 6.3 compares the transparency of nickel micro-EMTE and ITO-PEN before and after the deposition of PtNPs. Initially, the nickel micro-EMTE has a lower optical transparency (74% at 550 nm) than that of ITO-PEN (84% at 550 nm) substrate. However,

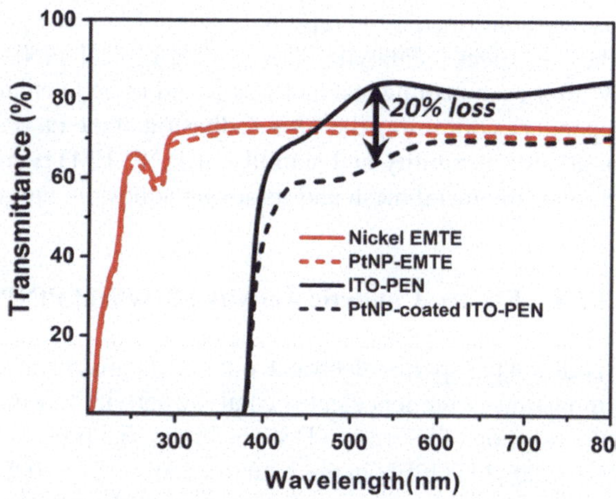

Fig. 6.3 Optical transmittance comparison of the-nickel micro-EMTE and ITO-PEN before and after PtNPs coating

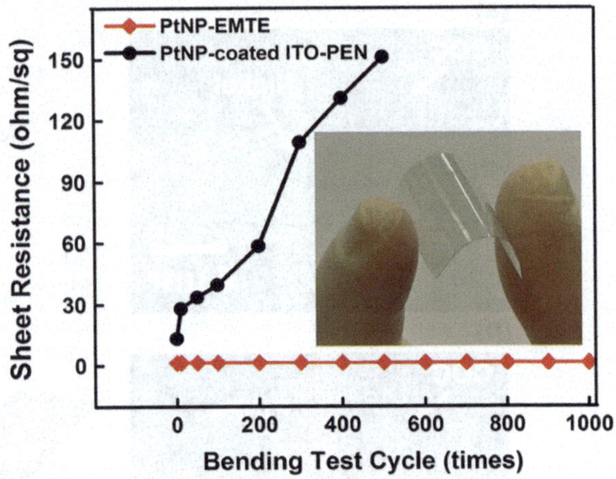

Fig. 6.4 Plot of variations in sheet resistance versus the number of cycles of the nickel micro-EMTE and ITO-PEN after repeated bending (compressive loading) to radii of 5 mm

after the PtNP deposition there is only 2% loss in optical transparency as compared to approximately 20% loss in a typical PtNP-coated ITO-PEN substrate. It is due to the fact that in pulsed electro-deposition, PtNPs are grown only on the surface of the conductive mesh, therefore does not significantly disturb the optical transparency of the nickel micro-EMTE, while in case of ITO-PEN, the PTNPs cover the whole substrate, resulting in significant loss of optical transparency. This in situ deposition of PtNPs results in overall higher optical transmittance of our hybrid PtNP-EMTE.

PtNP-EMTEs also exhibit superior mechanical stability under cyclic bending stress as compared to PtNP-coated ITO-PEN. Figure 6.4 presents the variation in sheet resistance as a function of the number of cycles for repeated compressive bending to a radius of 5 mm. The results clearly indicate that there is no noticeable change in sheet resistance (R_{Sh} = 1.32 Ω/sq) for up to 1000 bending cycles. In contrast, severe degradation in conductivity is observed in PtNP-coated ITO-PEN films after just a few bending cycles. The mechanical stability of the PtNP-EMTE was also confirmed using repeated peeling test using polypropylene tape with acrylic adhesive. Electro-catalytic activity of a typical PtNP-EMTE was measured using CV after each peeling test and was found to be unchanged after 3 cycles (Fig. 6.17, Appendix), confirming the strong adhesion of the PtNPs on nickel micro-EMTE. The remarkable flexibility and stability of PtNP-EMTE are attributed to the embedded nature of the metal mesh and its strong adhesion with the electrodeposited PtNPs.

6.2.2.3 Electro-Catalytic Activity of Hybrid PtNP-EMTE

In addition to superior electrical, optical and mechanical performances, PtNP-EMTE demonstrates excellent electro-catalytic activity toward I^-/I_3^- redox reaction, which is the reaction of interest in DSSCs. Figure 6.5 presents the CVs of PtNP-EMTE and PtNP-coated ITOPEN in the scan range of −0.4 to 0.4 V, showing the pair of anodic and cathodic peaks along with peak current density (I_{peak}), which is typically used to

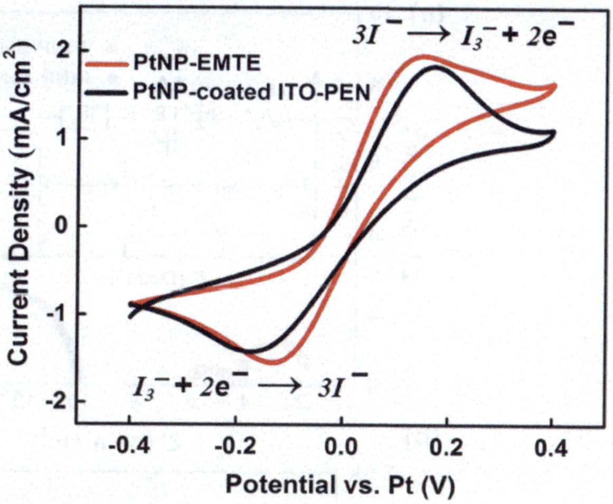

Fig. 6.5 Cyclic voltammograms of the PtNP-EMTE and PtNP-coated ITO-PEN

evaluate the electrochemical properties of the CEs [61, 62]. It is obvious from the data that although the effective area of PtNP-EMTE is only 10% of that on ITO-PEN fully covered with PtNPs, higher I_{peak} is observed for PtNP-EMTE, revealing its higher electro-catalytic activity of I^-/I_3^- redox reaction. This higher I_{peak} is attributed to the increased active surface area of PtNP clusters formed during pulse electrodeposition. The X-ray diffraction (XRD) characterization (Fig. 6.18, Appendix) further confirms this by revealing the high intensity of two low-index Pt facets, Pt (111) and Pt (110) planes, which are catalytically active to I_3^- reduction [63–65] according to the Joint Committee on Powder Diffraction standards (JCPDS, PDF no. 04-0802). The superior performances of hybrid PtNP-EMTE indicate that the integrated use of metal mesh and PtNPs is the ideal solution for CEs of bifacial flexible DSSCs as it takes advantage of the high electro-catalytic activity of PtNPs together with the excellent optical, electrical and mechanical qualities of metal-meshes.

EIS measurements using a symmetric cell were further carried out to understand the superior performance of our PtNP-EMTE CE, as described in details in Sect. 6.2). Nyquist plot shown in Fig. 6.6a manifests the interfacial processes with three clearly distinguishable impedance features from high to low frequency representing series resistance (R_S), charge transfer resistance (R_{CT}) at the interface between the electrolyte and the CE and Nernst diffusion resistance (R_D) of I^-/I_3^- in the electrolyte respectively. The values of equivalent-circuit elements corresponding to the best-fit curves in Fig. 6.6a are summarized in Table 6.1. Due to the superior electrical conductivity, the Rs (0.80 Ω-cm^2) for PtNP-EMTE is much lower than that of PtNP-coated ITO-PEN (5.05 Ω-cm^2). Figure 6.6b shows the R_{CT} (corresponding best-fit Nyquist plots are shown in Fig. 6.19 (Appendix)) of PtNP-EMTEs (active area = 0.36 cm^2) as a function of pulsed-electrodeposition time. It is evident from the graph that with increase of pulsed electro-deposition time, the density of PtNPs increases, as shown in the inset SEM pictures in Fig. 6.6b, leading to a significant decrease of R_{CT}. Once reached a saturated value (0.21 mg cm^{-2}), additional deposition of PtNPs

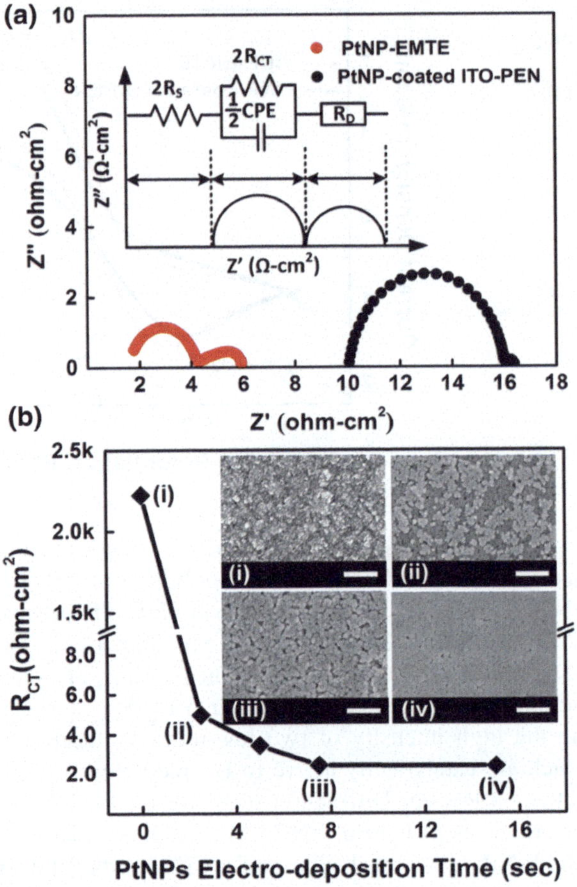

Fig. 6.6 Electrochemical characterizations. **a** Nyquist plot of the dummy cells with corresponding equivalent circuit; and **b** equivalent charge transfer resistance of PtNP-EMTE as a function of PtNPs electrodeposition time; the insets show the SEM images (all scalebars = 300 nm) of the corresponding samples

Table 6.1 Electrical properties from symmetric dummy cell

Counter electrodes	R_{Sh} (Ω/sq)	R_S (Ω cm²)	R_{CT} (Ω cm²)	C_{PE} (μF/cm²)	R_D (Ω cm²)
PtNP-EMTE	1.32	0.8	1.21	8.2	2.15
PtNP-coated ITOPEN	13.3	5.05	2.93	56	0.43

has no significant effect on the R_{CT} and therefore an optimized loading is achieved. At the optimized loading, PtNP-EMTE exhibited lower R_{CT} of 1.21 Ω-cm^2 compared to PtNP-coated ITO-PEN (2.93 Ω-cm^2). Furthermore, the significantly better electro-catalytic activity of PtNP-EMTE is also confirmed by the Bode plot presented in Fig. 6.20 (Appendix). The characteristic frequency corresponding to R_{CT} of PtNP-EMTE shifts to a much higher value of 43.3 kHz compared with 2.1 kHz of PtNP-coated ITO-PEN, indicating the improved electro-catalytic activity with a faster reaction response of the ions toward PtNPs.

6.2.2.4 Flexible Unifacial DSSCs Using PtNP-EMTE as Counter Electrode

The superior electrical, optical, mechanical and electrochemical properties of our PtNP-EMTE make it an excellent replacement of conventional ITO-PEN CE in flexible DSSCs. Following the procedure similar to the one presented in Fig. 6.1, flexible DSSCs using PtNP-EMTE as the CE and titanium foil (Ti-foil) as the photo-anode are fabricated to compare with control samples that use PtNP-coated ITO-PEN as the CE. Figure 6.7a shows the schematic diagram and photograph of the device. Since this architecture uses the opaque Ti-foil as photo-anode, the cell can be illuminated by light only from the rear-side and therefore is a flexible unifacial DSSC. Current density-voltage (J-V) characteristic of both DSSCs using PtNP-EMTE and ITO-PEN as CEs was characterized under simulated AM 1.5G solar illumination of 100 mW cm^{-2} and demonstrated in Fig. 6.7b. PtNP-EMTE DSSC device achieved short-circuit current (J_{sc}), open-circuit voltage (V_{oc}) and fill factor (FF) of 13.9 mA cm^{-2}, 0.79 V, and 67.3%, respectively, exhibiting an excellent PCE of 6.91%, which is much higher than that of 4.57% achieved on the ITO-PEN DSSC. We attribute the improved performance of the PtNP-EMTE DSSC to the higher transmission, lower sheet resistance, and satisfactory electro-catalytic activity of this hybrid CE.

6.2.2.5 Flexible Bifacial DSSCs Using PtNP-EMTE as Counter Electrode

Using the LEIT fabricated PtNP-EMTE as CE and ITO-PEN as photo-anode, flexible bifacial DSSCs were demonstrated and compared with control samples of DSSCs using ITO-PEN for both CE and photo-anode, as shown in Fig. 6.8. Figure 6.8a represents the schematic diagram and photograph of fabricated flexible bifacial DSSCs. These DSSCs can be illuminated by light from both sides and their J-V curves and key performance are compared in Fig. 6.8b and Table 6.2. Under front (rear) illumination, the device with PtNP-EMTE CE exhibited a J_{sc} of 11.63 (9.98) mA cm^{-2}, a V_{oc} of 0.70 (0.70) V, and a FF of 68.1 (67.1) %, resulting in a PCE of 5.52 (4.71) %, which is clearly improved from that of 4.31 (2.11) % PCE as achieved on the DSSC with PtNP-coated ITO-PEN CE. The bracketed data are for the device illuminated

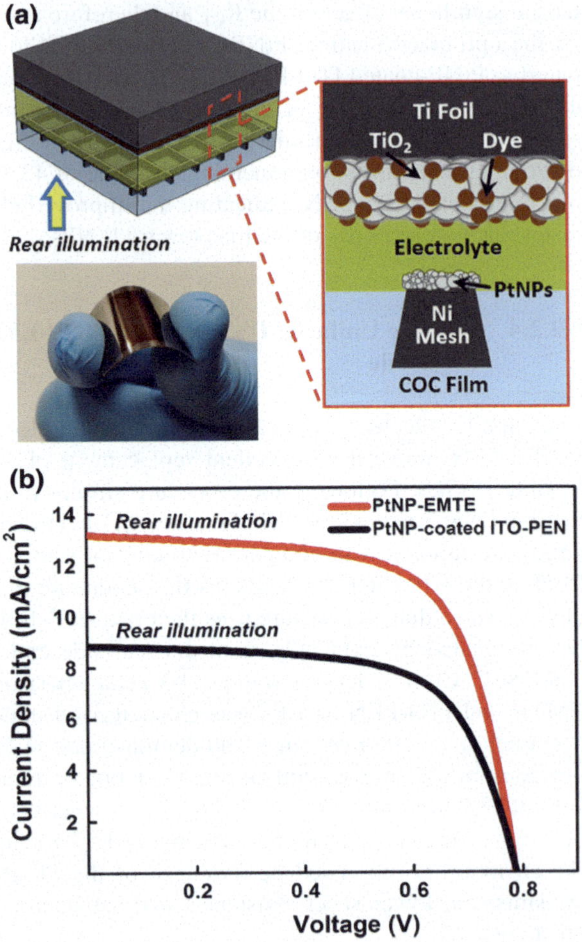

Fig. 6.7 Flexible unifacial DSSCs using PtNP-EMTE as CE and Ti-foil as photo-anode. **a** Schematic diagram and photograph of the device. **b** Comparison of J–V curves under rear illumination

from the CE side. This excellent exhibition of PCE under rear illumination, which is more than 85% that under the front illumination, can be attributed to the synergetic effects of the PtNP-EMTE outperforming the corresponding PtNP-coated ITO-PEN.

Also, flexible bifacial DSSCs using the TEIT fabricated PtNP-EMTE as CE and ITO-PEN as photo-anode, were made-up and characterized. Figure 6.9 displays the J–V characteristic of TEIT based PtNP-EMTE DSSC device under simulated AM 1.5G solar illumination of 100 mW cm^{-2}. Under front illumination, this DSSC demonstrated a Jsc of 11.10 mA cm^{-2}, a Voc of 0.72 V, and a FF of 64.9%, resulting in a PCE of 5.21%. Under rear illumination, this cell accomplished a Jsc of 9.25 mA cm^{-2}, a Voc of 0.74 V, and a FF of 65.4%, giving a PCE of 4.53%. The PCE under rear illumination of this bifacial DSSC is also above 85% that of under the front illumination which further confirms the better performance of the micro-EMTE based CE.

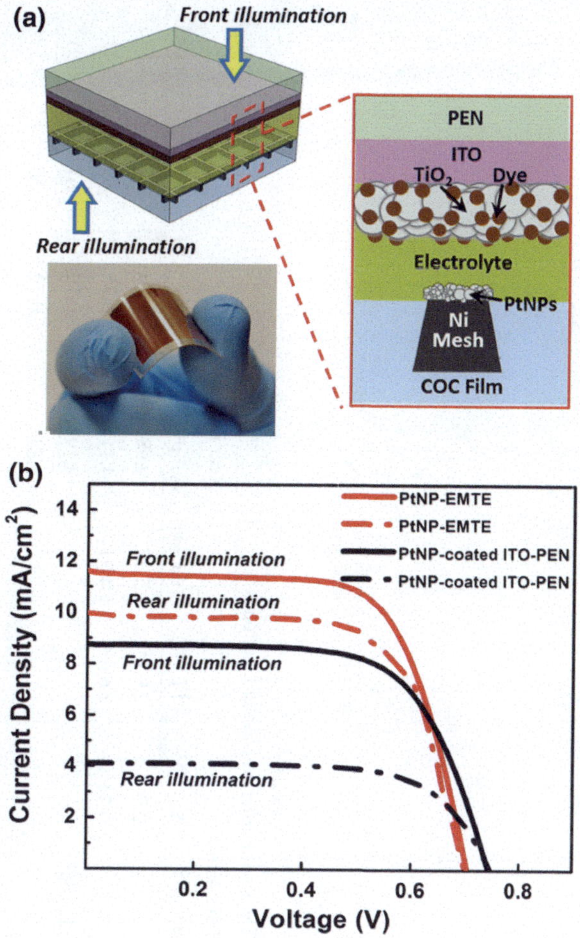

Fig. 6.8 Flexible bifacial DSSCs using LEIT fabricated PtNP-EMTE as CE and ITO-PEN as photo-anode. **a** Schematic diagram and photograph of the device. **b** Comparison of J–V curves for both front-illumination and rear-illumination

From the viewpoint of the daily use of the flexible DSSCs, the mechanical flexibility and durability under bending stress are of great importance concerning practical applications, such as portable and wearable electronics. Bending tests, taking into account the bending radii, were carried out to evaluate the device stability against mechanical bending. The bifacial flexible DSSCs were bent to four different radii of curvature (r = 2, 1.5, 1 and 0.5 cm) and the J–V curves were measured in curved form (Fig. 6.21, Appendix) under both front and rear illumination. Figure 6.10a displays the J–V curves under front illumination, where the calculated PCE shows less than 4% decrease in efficiency till 1 cm bending radius, accompanying the well-maintained key photovoltaic parameters such as J_{SC}, V_{OC} and FF (Fig. 6.22, Appendix). Such high bending durability is very competitive to the state-of-the-art in flexible DSSCs. However, by bending the device to 0.5 cm radius, there is obvious decrease in the

Table 6.2 Photovoltaic performance of the flexible DSSCs with different counter electrodes

Photo-anode	Counter electrode	Front illumination				Rear illumination			
		J_{sc} (mA/cm^2)	V_{oc} (V)	FF (%)	η (%)	J_{sc} (mA/cm^2)	V_{oc} (V)	FF (%)	η (%)
ITO-PEN	PtNP-EMTE	11.63	0.70	68.1	5.67	9.98	0.70	67.1	4.87
	PtNP-coated ITO-PEN	8.76	0.74	66.0	4.31	4.11	0.75	69.0	2.11
Ti-Foil	PtNP-EMTE	Not applicable				13.19	0.79	67.3	6.91
	PtNP-coated ITO-PEN					8.84	0.79	66.1	4.57

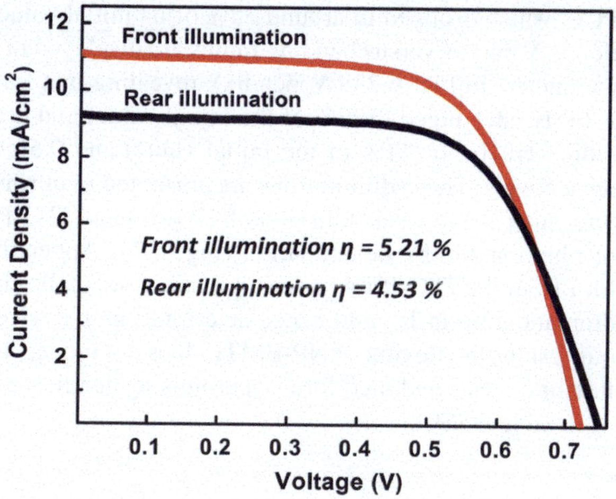

Fig. 6.9 J–V curves (front-illumination and rear-illumination) of flexible bifacial DSSCs using TEIT fabricated PtNP-EMTE as CE and ITO-PEN as photo-anode

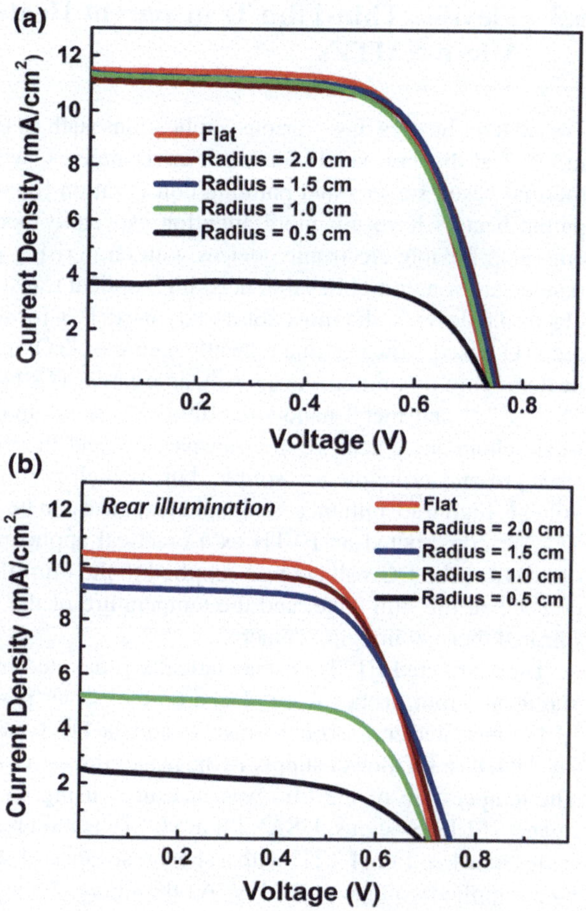

Fig. 6.10 Flexibility tests on bifacial DSSCs using PtNP-EMTE as CE and ITO-PEN as photo-anode. **a** J–V curves measured at various bending radius under front illumination. **b** J–V curves measured at various bending radius under rear illumination

PCE, which dropped to around 35% of its initial value. Similarly for rear illumination, J-V curves versus bending radius are displayed in Fig. 6.9b and its photovoltaic parameters in Fig. 6.23 (Appendix), revealing that for 2 and 1.5 cm radii, the PCE is fairly unchanged (>90% of the initial value) and drastically drops when bending radii were 1 cm (51% of the initial value) and 0.5 cm (18% of the initial value), respectively. These diminutions are attributed to intrinsic issues of this device structure, such as the weak adhesion of TiO_2 film on ITO-PEN, which caused the failure of photo-anode by delamination (Fig. 6.24, Appendix) of the dye-adsorbed TiO_2 film from the ITO-PEN under high stress at smaller bending radii, resulting in the dramatic drop in J_{SC} and hence deterioration in device performances. According to our results, the flexible PtNP-EMTE does not induce noticeable degradation in these flexibility tests and therefore is a promising candidate for large-scale manufacturing of flexible DSSCs.

6.3 Flexible Thin-Film Transparent Heaters Using Micro-EMTEs

Transparent heaters have various applications such as in outdoor panel displays, liquid crystal display panels for harsh environments, window defrosters, periscopes, thermal-based sensors and painting conservation [66–69]. Recently, flexible transparent heaters have got more attention especially because of the development of emerging flexible electronic devices with large rising market for smart windows in residences or automobiles that need high optical transparency [70]. Similar to other electronic devices, the most commonly used TCs for traditional transparent heaters are ITO-based, however due to brittle nature of ITO, its usage in flexible transparent heaters is highly limited. As a result, alternative TCs based on carbon materials, [67, 68, 71, 72] and metal nanowires [69, 73] are the main subject of recent research to simultaneously achieve the transparency and flexibility in FTTH's. An FTTH's structure and principle are simple, but its high-performance operation with a low voltage, high transmittance, and rapid response can be achieved only with a superior TC. We constructed an FTTH as a practical application of our micro-EMTEs. In our device, the DC voltage was supplied to the thin-film heater through silver paste contacts at the film edge, and the temperature of the film was monitored using an infrared thermal imaging camera.

The fabricated FTTH is schematically illustrated in Fig. 6.11a. During the measurements, four probes were placed on two silver paste-covered edges of a square LEIT fabricated copper micro-EMTE, and the DC voltage was supplied by the Keithley 2400 as a DC power supply to the heater through a silver contact at the film edge. The temperature of the film was measured using an FLIR ONE infrared thermal imager (FLIR Systems, USA). Figure 6.11b is a representative steady-state thermal image of a 1×1 cm FTTH with a sheet resistance of 0.3 Ω/sq and which is powered by an applied voltage of 0.21 V. As the figure shows, the temperature distribution

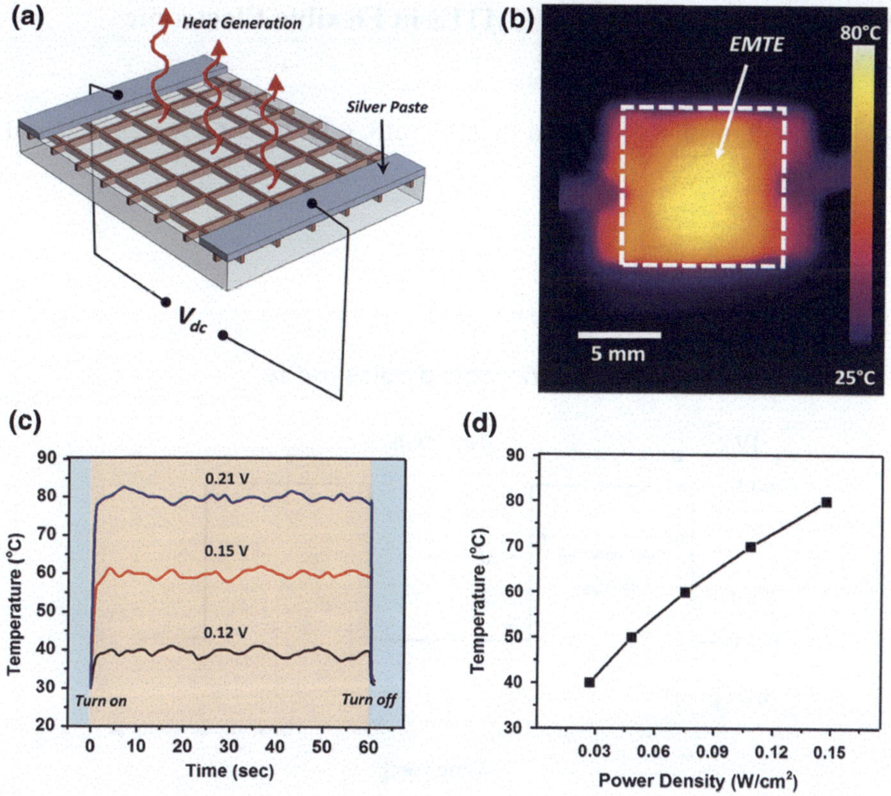

Fig. 6.11 Demonstration of micro-EMTE based FTTH. **a** Schematic illustration. **b** Infrared thermal image showing uniform temperature distribution over the entire FTTH during operation. **c** Plot of the temperature at FTTH center versus time at various applied voltages. **d** Plot of the FTTH temperature at steady state versus calculated input power density

is rather homogeneous; this is due to the excellent thermal and electrical conductivity of our micro-EMTE. The time-dependent temperature change of the heater is experimentally measured under various applied voltages from 0.12 to 0.21 V, as displayed in Fig. 6.11c. Regardless of the input voltages, the steady-state temperature of the FTTH was reached within 2 s, demonstrating the rapid response of the device. The center temperature of the FTTH reached 80 °C under 0.21 V and the power density was calculated to be approximately 0.15 W/cm^2 (Fig. 6.10d), confirming its operation at low input voltages and with a low power density requirement, which is mainly because of the low sheet resistance of our micro-EMTE. Accurate control of the FTTH temperature was achieved by adjusting the supplied voltage, as demonstrated in Fig. 6.11c. Compared with most published results, [46, 66–69, 74–77] a lower power density requirement (<0.15 W/cm^2), a rapid response (<2 s), and a lower operating voltage (<1 V) make our EMTE-based FTTH to be a unique device for broad applications.

Appendix: Applications of EMTEs in Flexible Electronic Devices

See Figs. 6.12, 6.13, 6.14, 6.15, 6.16, 6.17, 6.18, 6.19, 6.20, 6.21, 6.22, 6.23 and 6.24.

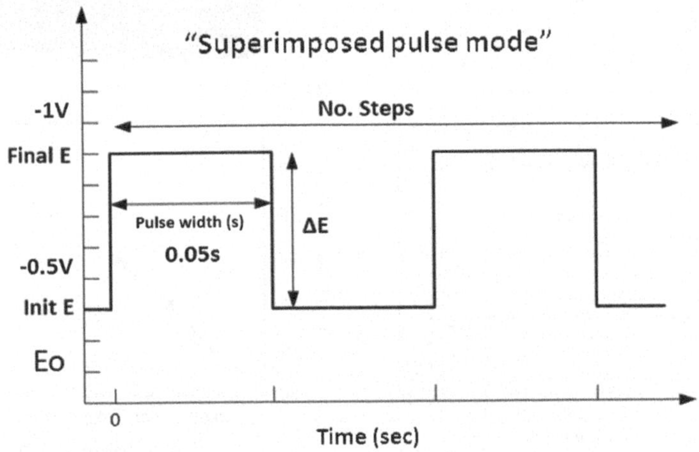

Fig. 6.12 Schematic illustration of the profile of pulsed waveform

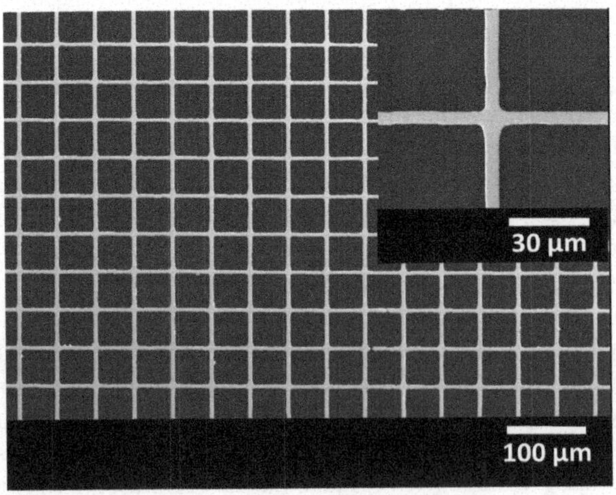

Fig. 6.13 SEM images of nickel micro-EMTE on flexible COC film

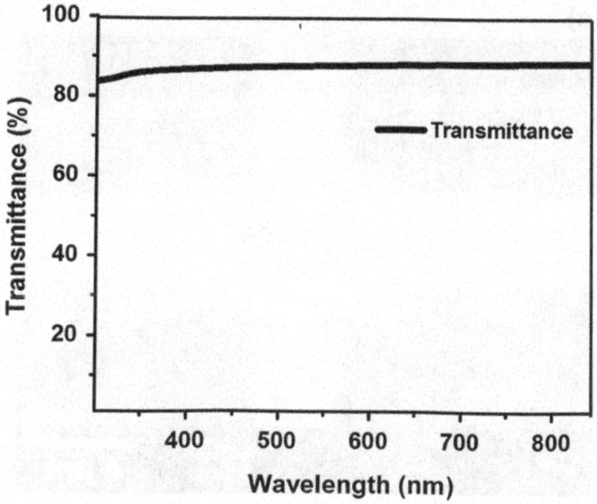

Fig. 6.14 Optical transparency of blank COC film in UV and visible wavelength range

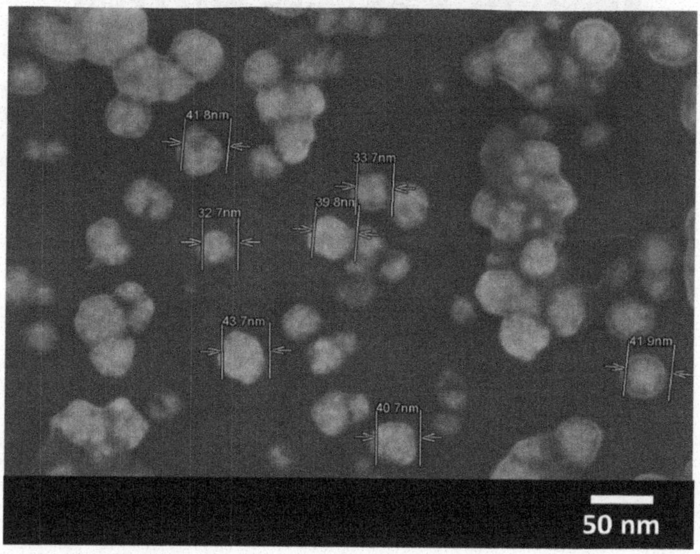

Fig. 6.15 SEM image showing the size of the PtNPs deposited by the pulsed electrodeposition technique

(a)

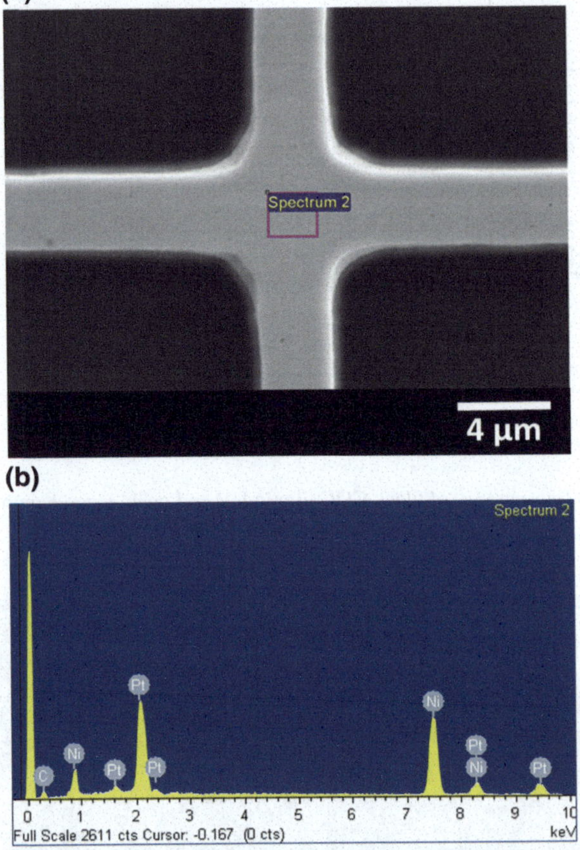

(c)

Element	Weight%	Atomic%
C K	8.72	40.41
Ni K	50.63	48.00
Pt M	40.65	11.60
Totals	100.00	

Fig. 6.16 SEM-EDS analysis of mesh lines on a PtNP-EMTE. **a** SEM micrograph of PtNP coated nickel micro-EMTE; **b** EDS spectrum of the corresponding boxed area in (**a**); and **c** elemental quantification table at the corresponding boxed area (in Figure a)

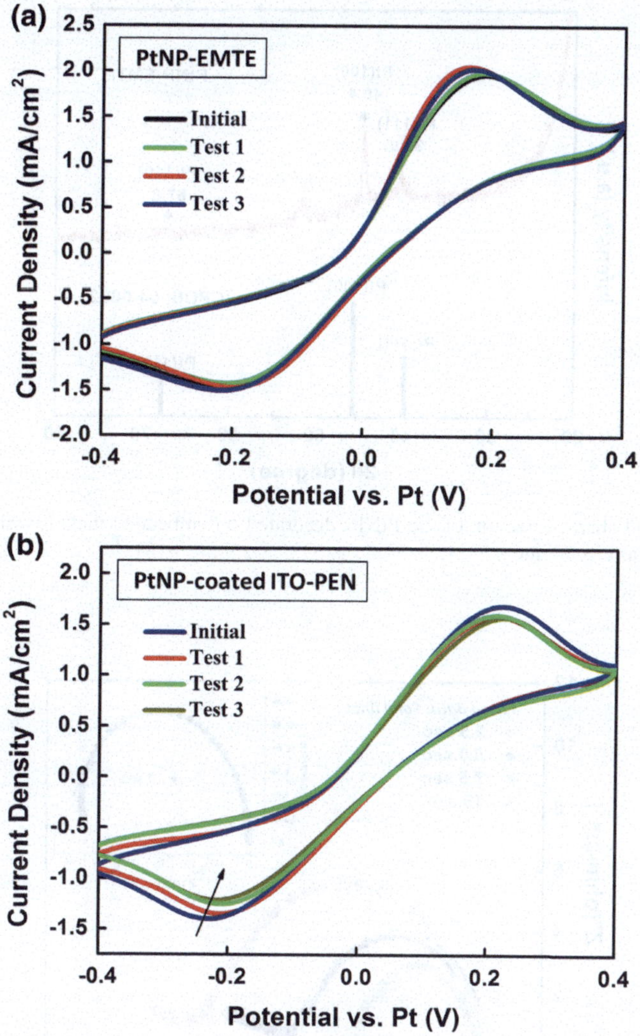

Fig. 6.17 Variations in CVs of CEs after the repeated adhesive tape tests on **a** PtNP-EMTE and **b** PtNP-coated ITO-PEN

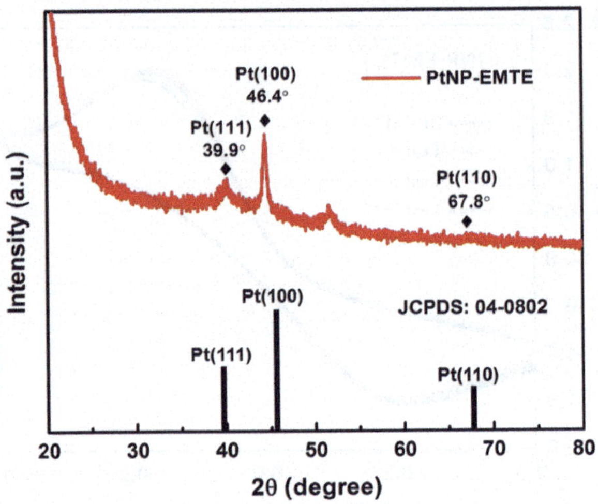

Fig. 6.18 XRD characterization of the PtNPs decorated on embedded nickel mesh using puled electro-deposition technique

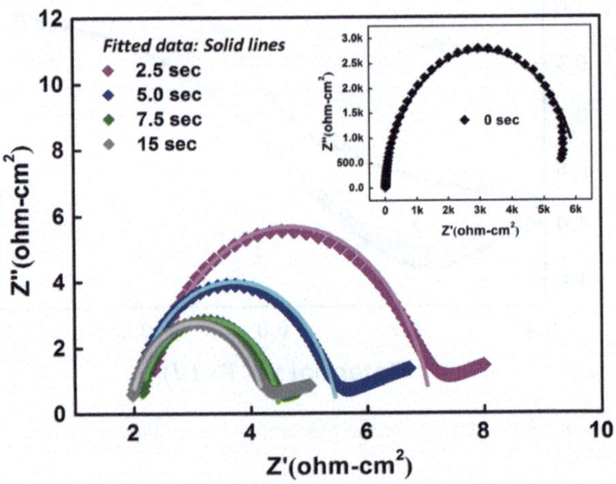

Fig. 6.19 Nyquist plots of the dummy cells utilizing PtNP-EMTEs prepared at different pulsed electrodeposition time

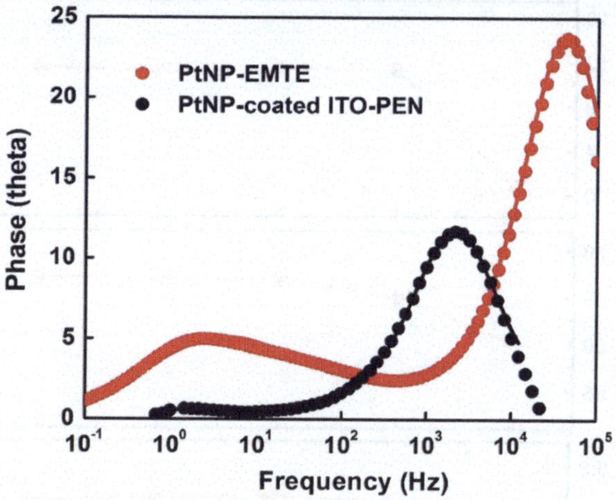

Fig. 6.20 Bode plots of the PtNP-EMTE and PtNP-coated ITO-PEN

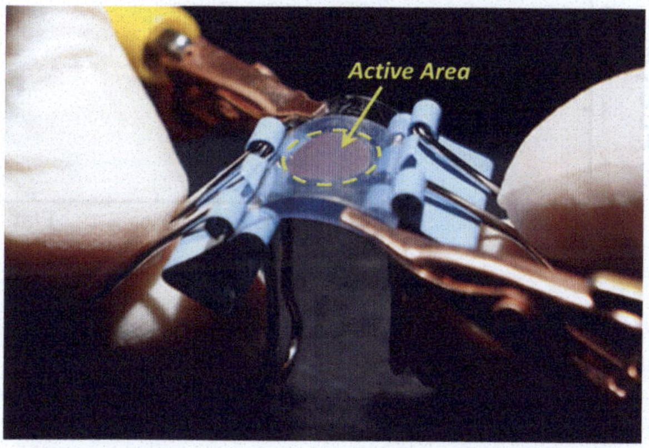

Fig. 6.21 Photograph of PtNP-EMTE based bifacial flexible DSSC in bended form

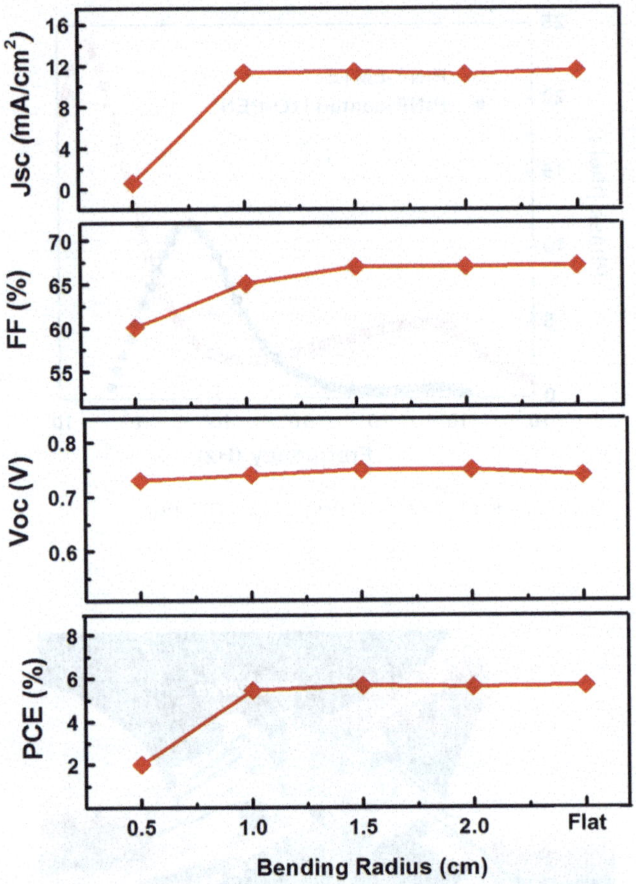

Fig. 6.22 Photovoltaic performance parameters of a bifacial flexible DSSC as function of bending radius by subjecting it to compressive loading and illuminated from front side

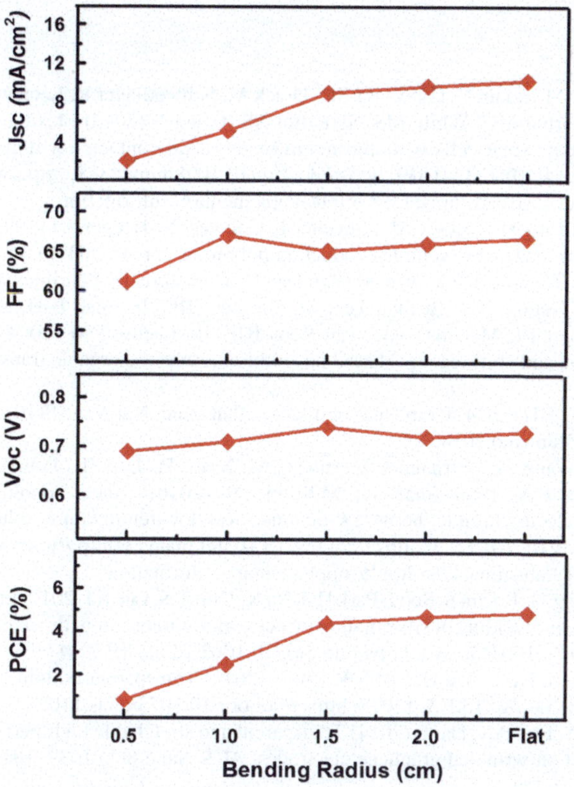

Fig. 6.23 Photovoltaic performance parameters of a bifacial flexible DSSC as function of bending radius by subjecting it to tensile loading and illuminated from rear side

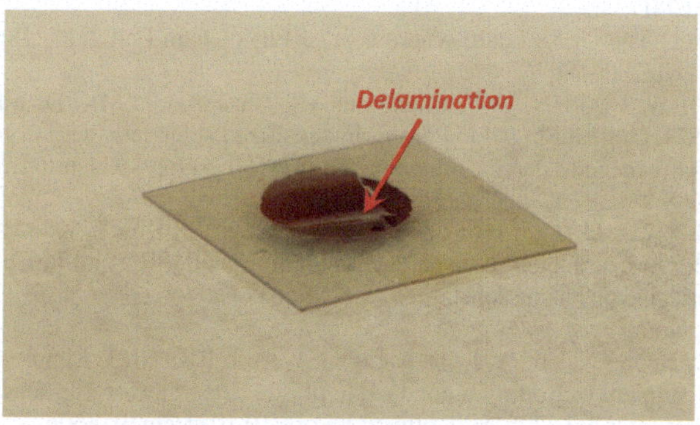

Fig. 6.24 Failure of photo-anode by the delamination of the dye-adsorbed TiO_2 film from the ITO-PEN substrate

References

1. Kaltenbrunner M, Adam G, Glowacki ED, Drack M, Schwodiauer R, Leonat L, Apaydin DH, Groiss H, Scharber MC, White MS, Sariciftci NS, Bauer S (2015) Flexible high power-per-weight perovskite solar cells with chromium oxide-metal contacts for improved stability in air. Nat Mater 14(10):1032–1039. https://doi.org/10.1038/nmat4388, http://www.nature.com/nmat/journal/v14/n10/abs/nmat4388.html#supplementary-information
2. Gustafsson G, Cao Y, Treacy GM, Klavetter F, Colaneri N, Heeger AJ (1992) Flexible light-emitting diodes made from soluble conducting polymers. Nat 357(6378):477–479
3. Gelinck GH, Huitema HEA, van Veenendaal E, Cantatore E, Schrijnemakers L, van der Putten JBPH, Geuns TCT, Beenhakkers M, Giesbers JB, Huisman B-H, Meijer EJ, Benito EM, Touwslager FJ, Marsman AW, van Rens BJE, de Leeuw DM (2004) Flexible active-matrix displays and shift registers based on solution-processed organic transistors. Nat Mater 3(2):106–110
4. Ahn J-H, Hong BH (2014) Graphene for displays that bend. Nat Nano 9(10):737–738. https://doi.org/10.1038/nnano.2014.226
5. Llordes A, Wang Y, Fernandez-Martinez A, Xiao P, Lee T, Poulain A, Zandi O, Saez Cabezas CA, Henkelman G, Milliron DJ (2016) Linear topology inamorphous metal oxide electrochromic networks obtained via low-temperature solution processing. Nat Mater 15(12):1267–1273. https://doi.org/10.1038/nmat4734http://www.nature.com/nmat/journal/v15/n12/abs/nmat4734.html#supplementary-information
6. Park JH, Hwang,G-T, Kim S, Seo J, Park H-J, Yu K, Kim T-S, Lee KJ (2017) Flash-induced self-limited plasmonic welding of silver nanowire network for transparent flexible energy harvester. Adv Mater 29(5):1603473-n/a. https://doi.org/10.1002/adma.201603473
7. Yang Y, Jeong S, Hu L, Wu H, Lee SW, Cui Y (2011) Transparent lithium-ion batteries. Proc Natl Acad Sci 108(32):13013–13018. https://doi.org/10.1073/pnas.1102873108
8. Chen T, Xue Y, Roy AK, Dai L (2014) Transparent and stretchable high-performance supercapacitors based on wrinkled graphene electrodes. ACS Nano 8(1):1039–1046. https://doi.org/10.1021/nn405939w
9. Gupta R, Rao KDM, Kiruthika S, Kulkarni GU (2016) Visibly Transparent Heaters. ACS Appl Mater Interfaces 8(20):12559–12575. https://doi.org/10.1021/acsami.5b11026
10. Li Y, Lee D-K, Kim JY, Kim B, Park N-G, Kim K, Shin J-H, Choi I-S, Ko MJ (2012) Highly durable and flexible dye-sensitized solar cells fabricatedon plastic substrates: PVDF-nanofiber-reinforced TiO2 photoelectrodes. Energy Environ Sci 5(10):8950–8957. https://doi.org/10.1039/C2EE21674D
11. Peter LM (2011) The Grätzel cell: where next? J Phys Chem Lett 2(15):1861–1867. https://doi.org/10.1021/jz200668q
12. Yella A, Lee H-W, Tsao HN, Yi C, Chandiran AK, Nazeeruddin MK, Diau EW-G, Yeh C-Y, Zakeeruddin SM, Grätzel M (2011) Porphyrin-sensitized solar cells with Cobalt (ii/iii)–Based redox electrolyte exceed 12 percent efficiency. Sci 334(6056):629–634. https://doi.org/10.1126/science.1209688
13. Yoo K, Kim J-Y, Lee JA, Kim JS, Lee D-K, Kim K, Kim JY, Kim B, Kim H, Kim WM, Kim JH, Ko MJ (2015) Completely transparent conducting oxide-free and flexible dye-sensitized solar cells fabricated on plastic substrates. ACS Nano 9(4):3760–3771. https://doi.org/10.1021/acsnano.5b01346
14. Gong Y, Li C, Huang X, Luo Y, Li D, Meng Q, Iversen BB (2013) Simple method for manufacturing pt counter electrodes on conductive plastic substratesfor dye-sensitized solar cells. ACS App Mater Inter 5(3):795–800. https://doi.org/10.1021/am302360g
15. Hübner A, Aberle AG, Hezel R (1997) Novel cost-effective bifacial silicon solar cells with 19.4% front and 18.1% rear efficiency. App Phys Lett 70(8):1008–1010. https://doi.org/10.1063/1.118466
16. Ito S, Zakeeruddin SM, Comte P, Liska P, Kuang D, Gratzel M (2008) Bifacial Dye-sensitized Solar Cells Based on an Ionic Liquid Electrolyte. Nat Photon 2(11):693–698

17. Song D, Li M, Li Y, Zhao X, Jiang B, Jiang Y (2014) Highly transparent and efficient counter electrode using sio2/pedot–pss composite for bifacial dye-sensitized solar cells. ACS Appl Mater Inter 6(10):7126–7132. https://doi.org/10.1021/am500082x
18. Wu J, Li Y, Tang Q, Yue G, Lin J, Huang M, Meng L (2014) Bifacial dye-sensitized solar cells: a strategy to enhance overall efficiency based on transparentpolyaniline electrode. Sci Rep 4:4028. https://doi.org/10.1038/srep04028, http://www.nature.com/articles/srep04028#supplementary-information
19. Hezel R (2003) Novel applications of bifacial solar cells. Progress in Photovoltaics: Res App 11(8):549–556. https://doi.org/10.1002/pip.510
20. Bu C, Liu Y, Yu Z, You S, Huang N, Liang L, Zhao X-Z (2013) Highly transparent carbon counter electrode prepared via an in situ carbonization method forbifacial dye-sensitized solar cells. ACS Appl Mater Inter 5(15):7432–7438. https://doi.org/10.1021/am4017472
21. Yun S, Hagfeldt A, Ma T (2014) Pt-free counter electrode for dye-sensitized solar cells with high efficiency. Adv Mater 26(36):6210–6237. https://doi.org/10.1002/adma.201402056
22. Tai Q, Chen B, Guo F, Xu S, Hu H, Sebo B, Zhao X-Z (2011) In Situ prepared transparent polyaniline electrode and its application in bifacial dye-sensitized solar cells. ACS Nano 5(5):3795–3799. https://doi.org/10.1021/nn200133g
23. Balasingam SK, Kang MG, Jun Y (2013) Metal substrate based electrodes for flexible dye-sensitized solar cells: fabrication methods, progress and challenges. Chem Commun 49(98):11457–11475. https://doi.org/10.1039/C3CC46224B
24. Lin L-Y, Lee C-P, Vittal R, Ho K-C (2010) Selective conditions for the fabrication of a flexible dye-sensitized solar cell with Ti/TiO2 photoanode. J Power Sources 195(13):4344–4349. https://doi.org/10.1016/j.jpowsour.2010.01.031
25. Wang Y, Zhao C, Qin D, Wu M, Liu W, Ma T (2012) Transparent flexible Pt counter electrodes for high performance dye-sensitized solar cells. J Mater Chem 22(41):22155–22159. https://doi.org/10.1039/C2JM35348B
26. Zardetto V, Di Giacomo F, Garcia-Alonso D, Keuning W, Creatore M, Mazzuca C, Reale A, Di Carlo A, Brown TM (2013) Fully plastic dye solar cell devices by low-temperature UV-irradiation of both the mesoporous TiO2 photo- and platinized counter-electrodes. Adv Energy Mater 3(10):1292–1298. https://doi.org/10.1002/aenm.201300101
27. Hauch A, Georg A (2001) Diffusion in the electrolyte and charge-transfer reaction at the platinum electrode in dye-sensitized solar cells. Electrochimica Acta 46(22):3457–3466. https://doi.org/10.1016/S0013-4686(01)00540-0
28. Hashmi G, Miettunen K, Peltola T, Halme J, Asghar I, Aitola K, Toivola M, Lund P (2011) Review of materials and manufacturing options for large area flexible dye solar cells. Renew Sustain Energy Rev 15(8):3717–3732. https://doi.org/10.1016/j.rser.2011.06.004
29. Lee K-M, Hsu C-Y, Chen P-Y, Ikegami M, Miyasaka T, Ho K-C (2009) Highly Porous PProDOT-Et2 film as counter electrode for plastic dye-sensitized solar cells. Phys Chem 11(18):3375–3379. https://doi.org/10.1039/B823011K
30. Pringle JM, Armel V, MacFarlane DR (2010) Electrodeposited PEDOT-on-plastic cathodes for dye-sensitized solar cells. Chem Commun 46(29):5367–5369. https://doi.org/10.1039/C0CC01400A
31. Miettunen K, Toivola M, Hashmi G, Salpakari J, Asghar I, Lund P (2011) A carbon gel catalyst layer for the roll-to-roll production of dye solar cells. Carbon 49(2):528–532. https://doi.org/10.1016/j.carbon.2010.09.052
32. Burschka J, Brault V, Ahmad S, Breau L, Nazeeruddin MK, Marsan B, Zakeeruddin SM, Gratzel M (2012) Influence of the counter electrode on the photovoltaic performance of dye-sensitized solar cells using a disulfide/thiolate redox electrolyte. Energy Environ Sci 5(3):6089–6097. https://doi.org/10.1039/C2EE03005E
33. Aitola K, Halme J, Feldt S, Lohse P, Borghei M, Kaskela A, Nasibulin AG, Kauppinen EI, Lund PD, Boschloo G, Hagfeldt A (2013) Highly catalyticcarbon nanotube counter electrode on plastic for dye solar cells utilizing cobalt-based redox mediator. Electrochimica Acta 111:206–209. https://doi.org/10.1016/j.electacta.2013.07.202

34. Qin Q, Zhang R (2013) A novel conical structure of polyaniline nanotubes synthesized on ito-pet conducting substrate by electrochemical method. Electrochimica Acta 89:726–731. https://doi.org/10.1016/j.electacta.2012.11.107
35. Hung K-H, Chan C-H, Wang H-W (2014) Flexible tco-free counter electrode for dye-sensitized solar cells using graphene nanosheets from a Ti–Ti(III) acid solution. Renew Energy 66:150–158. https://doi.org/10.1016/j.renene.2013.12.001
36. Hung K-H, Wang H-W (2014) A freeze-dried graphene counter electrode enhances the performance of dye-sensitized solar cells. Thin Solid Films 550:515–520. https://doi.org/10.1016/j.tsf.2013.10.129
37. Xu X, Yang W, Li Y, Tu Z, Wu S, Ma X, Yang F, Zhang L, Chen S, Wang A, Liu H, Richard P (2015) Heteroatom-doped graphene-like carbon filmsprepared by chemical vapour deposition for bifacial dye-sensitized solar cells. Chem Eng J 267:289–296. https://doi.org/10.1016/j.cej.2015.01.028
38. Yang W, Xu X, Tu Z, Li Z, You B, Li Y, Raj SI, Yang F, Zhang L, Chen S, Wang A (2015) Nitrogen plasma modified CVD grown graphene as counterelectrodes for bifacial dye-sensitized solar cells. Electrochimica Acta 173:715–720. https://doi.org/10.1016/j.electacta.2015.05.143
39. Chen L, Tan W, Zhang J, Zhou X, Zhang X, Lin Y (2010) Fabrication of high performance Pt counter electrodes on conductive plastic substrate for flexible dyesensitized solar cells. Electrochimica Acta 55(11):3721–3726. https://doi.org/10.1016/j.electacta.2010.01.108
40. Fang X, Ma T, Guan G, Akiyama M, Kida T, Abe E (2004) Effect of the thickness of the Pt film coated on a counter electrode on the performance of a dyesensitizedsolar cell. J Electroanal Chem 570(2):257–263. https://doi.org/10.1016/j.jelechem.2004.04.004
41. Fang X, Ma T, Akiyama M, Guan G, Tsunematsu S, Abe E (2005) Flexible counter electrodes based on metal sheet and polymer film for dye-sensitized solar cells. Thin Solid Films 472(1–2):242–245. https://doi.org/10.1016/j.tsf.2004.07.083
42. Garcia-Alonso D, Zardetto V, Mackus AJM, De Rossi F, Verheijen MA, Brown TM, Kessels WMM, Creatore M (2014) Atomic layer deposition of highly transparent platinum counter electrodes for metal/polymer flexible dye-sensitized solar cells. Adv Energy Mater 4(4):1300831–n/a. https://doi.org/10.1002/aenm.201300831
43. Yang L, Wu L, Wu M, Xin G, Lin H, Ma T (2010) High-efficiency flexible dye-sensitized solar cells fabricated by a novel friction-transfer technique. Electrochem Commun 12(7):1000–1003. https://doi.org/10.1016/j.elecom.2010.05.026
44. Fu N, Xiao X, Zhou X, Zhang J, Lin Y (2012) Electrodeposition of platinum on plastic substrates as counter electrodes for flexible dye-sensitized solar cells. J Phys Chem C 116(4):2850–2857. https://doi.org/10.1021/jp206676s
45. Hasin P, Alpuche-Aviles MA, Li Y, Wu Y (2009) Mesoporous Nb-Doped TiO2 as Pt support for counter electrode in dye-sensitized solar cells. J Phys Chem C 113(17):7456–7460. https://doi.org/10.1021/jp810678a
46. Hong S, Yeo J, Kim G, Kim D, Lee H, Kwon J, Lee H, Lee P, Ko SH (2013) Nonvacuum, maskless fabrication of a flexible metal grid transparentconductor by low-temperature selective laser sintering of nanoparticle ink. ACS Nano 7(6):5024–5031. https://doi.org/10.1021/nn400432z
47. Wu H, Kong D, Ruan Z, Hsu P-C, Wang S, Yu Z, Carney TJ, Hu L, Fan S, Cui Y (2013) A transparent electrode based on a metal nanotrough network. Nat Nano 8(6):421–425. https://doi.org/10.1038/nnano.2013.84, http://www.nature.com/nnano/journal/v8/n6/abs/nnano.2013.84.html#supplementary-information
48. Han B, Pei K, Huang Y, Zhang X, Rong Q, Lin Q, Guo Y, Sun T, Guo C, Carnahan D, Giersig M, Wang Y, Gao J, Ren Z, Kempa K (2014) Uniform Self-forming metallic network as a high-performance transparent conductive electrode. Adv Mater 26(6):873–877. https://doi.org/10.1002/adma.201302950
49. Khan A, Lee S, Jang T, Xiong Z, Zhang C, Tang J, Guo LJ, Li W-D (2016) High-performance flexible transparent electrode with an embedded metal mesh fabricated by cost-effective solution process. Small 12(22):3021–3030. https://doi.org/10.1002/smll.201600309

50. Huang J, Li G, Yang Y (2008) A Semi-transparent plastic solar cell fabricated by a lamination process. Adv Mater 20(3):415–419. https://doi.org/10.1002/adma.200701101
51. Kim A, Won Y, Woo K, Kim C-H, Moon J (2013) Highly transparent low resistance ZnO/Ag Nanowire/ZnO composite electrode for thin film solar cells. ACS Nano 7(2):1081–1091. https://doi.org/10.1021/nn305491x
52. Margulis GY, Christoforo MG, Lam D, Beiley ZM, Bowring AR, Bailie CD, Salleo A, McGehee MD (2013) Spray deposition of silver nanowire electrodes for semitransparent solid-state dye-sensitized solar cells. Adv Energy Mater 3(12):1657–1663. https://doi.org/10.1002/aenm.201300660
53. Bryant D, Greenwood P, Troughton J, Wijdekop M, Carnie M, Davies M, Wojciechowski K, Snaith HJ, Watson T, Worsley D (2014) A transparent conductive adhesive laminate electrode for high-efficiency organic-inorganic lead halide perovskite solar cells. Adv Mater 26(44):7499–7504. https://doi.org/10.1002/adma.201403939
54. Mao L, Chen Q, Li Y, Li Y, Cai J, Su W, Bai S, Jin Y, Ma C-Q, Cui Z, Chen L (2014) Flexible Silver Grid/PEDOT:PSS hybrid electrodes for large area inverted polymer solar cells. Nano Energy 10:259–267. https://doi.org/10.1016/j.nanoen.2014.09.007
55. Li Y, Meng L, Yang Y, Xu G, Hong Z, Chen Q, You J, Li G, Yang Y, Li Y (2016) High-efficiency robust perovskite solar cells on ultrathin flexible substrates. Nat Commun 7. https://doi.org/10.1038/ncomms10214
56. Servaites JD, Yeganeh S, Marks TJ, Ratner MA (2010) Efficiency enhancement in organic photovoltaic cells: consequences of optimizing series resistance. Adv Funct Mater 20(1):97–104. https://doi.org/10.1002/adfm.200901107
57. Rowell MW, McGehee MD (2011) Transparent electrode requirements for thin film solar cell modules. Energy Environ Sci 4(1):131–134. https://doi.org/10.1039/C0EE00373E
58. Khan A, Huang Y-T, Miyasaka T, Ikegami M, Feng S-P, Li W-D (2017) Solution-processed transparent nickel-mesh counter electrode with in-Situ electrodeposited platinum nanoparticles for full-plastic bifacial dye-sensitized solar cells. ACS Appl Mater Inter 9(9);8083–8091. https://doi.org/10.1021/acsami.6b14861
59. Khan A, Li W (2016) Transparent conductive films with embedded metal grids. Google Patents
60. Khan A, Lee S, Jang T, Xiong Z, Zhang C, Tang J, Guo LJ, Li W-D (2017) Scalable solution-processed fabrication strategy for high-performance. flexible, transparent electrodes with embedded metal mesh. 124:e56019. https://doi.org/10.3791/56019
61. Snaith HJ, Schmidt-Mende L (2007) Advances in liquid-electrolyte and solid-state dye-sensitized solar cells. Adv Mat 19(20):3187–3200. https://doi.org/10.1002/adma.200602903
62. Hardin BE, Snaith HJ, McGehee MD (2012) The renaissance of dye-sensitized solar cells. Nat Photon 6(3):162–169
63. Kolics A, Wieckowski A (2001) Adsorption of bisulfate and sulfate anions on a Pt(111) electrode. J Phys Chem B 105(13):2588–2595. https://doi.org/10.1021/jp003536f
64. Zhang H, Zhou W, Du Y, Yang P, Wang C (2010) One-step electrodeposition of platinum nanoflowers and their high efficient catalytic activity for methanolelectro-oxidation. Electrochem Commun 12(7):882–885. https://doi.org/10.1016/j.elecom.2010.04.011
65. Hsieh T-L, Chen H-W, Kung C-W, Wang C-C, Vittal R, Ho K-C (2012) A highly efficient dye-sensitized solar cell with a platinum nanoflowers counterelectrode. J Mater Chem 22(12):5550–5559. https://doi.org/10.1039/C2JM14623A
66. Yoon YH, Song JW, Kim D, Kim J, Park JK, Oh SK, Han CS (2007) Transparent film heater using single-walled carbon nanotubes. Adv Mater 19(23):4284–4287. https://doi.org/10.1002/adma.200701173
67. Kang J, Kim H, Kim KS, Lee S-K, Bae S, Ahn J-H, Kim Y-J, Choi J-B, Hong BH (2011) High-performance graphene-based transparent flexible heaters. Nano Lett 11(12):5154–5158. https://doi.org/10.1021/nl202311v
68. Sui D, Huang Y, Huang L, Liang J, Ma Y, Chen Y (2011) Flexible and transparent electrothermal film heaters based on graphene materials. Small 7(22):3186–3192. https://doi.org/10.1002/smll.201101305

69. Kim T, Kim YW, Lee HS, Kim H, Yang WS, Suh KS (2013) Uniformly Interconnected Silver-Nanowire Networks for Transparent Film Heaters. Adv Funct Mater 23(10):1250–1255. https://doi.org/10.1002/adfm.201202013
70. Kamalisarvestani M, Saidur R, Mekhilef S, Javadi FS (2013) Performance, materials and coating technologies of thermochromic thin films on smart windows. Renew Sustain Energy Rev 26:353–364. https://doi.org/10.1016/j.rser.2013.05.038
71. Liu L, Ma W, Zhang Z (2011) Macroscopic carbon nanotube assemblies: preparation, properties, and potential applications. Small 7(11):1504–1520. https://doi.org/10.1002/smll.201002198
72. Bae JJ, Lim SC, Han GH, Jo YW, Doung DL, Kim ES, Chae SJ, Huy TQ, Van Luan N, Lee YH (2012) Heat dissipation of transparent graphene defoggers. Adv Funct Mater 22(22):4819–4826. https://doi.org/10.1002/adfm.201201155
73. Yun S, Niu X, Yu Z, Hu W, Brochu P, Pei Q (2012) Compliant silver nanowire-polymer composite electrodes for bistable large strain actuation. Adv Mater 24(10):1321–1327. https://doi.org/10.1002/adma.201104101
74. Jang H-S, Jeon SK, Nahm SH (2011) The manufacture of a transparent film heater by spinning multi-walled carbon nanotubes. Carbon 49(1):111–116. https://doi.org/10.1016/j.carbon.2010.08.049
75. Celle C, Mayousse C, Moreau E, Basti H, Carella A, Simonato J-P (2012) Highly flexible transparent film heaters based on random networks of silver nanowires. Nano Res 5(6):427–433. https://doi.org/10.1007/s12274-012-0225-2
76. Gupta R, Rao KDM, Srivastava K, Kumar A, Kiruthika S, Kulkarni GU (2014) spray coating of crack templates for the fabrication of transparent conductors and heaters on flat and curved surfaces. ACS Appl Mater Interfaces 6(16):13688–13696. https://doi.org/10.1021/am503154z
77. Huang Q, Shen W, Fang X, Chen G, Guo J, Xu W, Tan R, Song W (2015) Highly flexible and transparent film heaters based on polyimide films embedded withsilver nanowires. RSC Adv 5(57):45836–45842. https://doi.org/10.1039/C5RA06529A

Chapter 7
Conclusions and Future Recommendations

Abstract This chapter summaries the key contributions of the dissertation and recommends directions for the future research.

7.1 Conclusions

This dissertation proposed and demonstrated a new type of embedded metal-mesh transparent electrodes with superior electrical, optical and mechanical properties. The structure of this novel EMTE allows the use of thick metal mesh for high conductivity without sacrificing surface smoothness. The embedded structure also improves the EMTE's mechanical stability under high bending stress and its chemical stability in an ambient environment. Moreover, it presents solution-processed fabrication approaches that potentially could be adapted for the high-throughput and low-cost manufacturing of the proposed transparent electrode. These fabrication processes have been demonstrated to be able to scale for a larger electrode area, narrower metal-mesh linewidths, and a wide range of materials. The approaches can be easily adapted to make flexible and even stretchable devices.

All the three major steps i.e. mesh patterning on conductive substrate, metal deposition into the mesh pattern and mesh transfer to flexible substrates; involve in our novel fabrication strategies are solution-based and do not require any expensive vacuum-based processing. These techniques are optimized to fabricate various prototype flexible micro-EMTEs and flexible nano-EMTEs with thick embedded metal-mesh, demonstrating excellent electrical and optical performances.

We applied EMTEs in flexible DSSCs and flexible FTTHs as practical applications. A new CE containing a micro-EMTE with catalytic PtNPs is developed for flexible bifacial DSSCs. This hybrid PtNP-decorated micro-EMTE based device demonstrates a remarkable PCE of 4.87% under rear-side illumination, which approaches 85% of the 5.67% power conversion efficiency (PCE) under front-side illumination, among the highest ratio in published results. Besides, an EMTE-based FTTH is constructed and characterized, demonstrating excellent performance compared with existing products. With superior performance and low-cost fabrication strategies, our EMTEs have broad applications in other flexible electronic devices.

A. Khan, *Novel Embedded Metal-mesh Transparent Electrodes*, Springer Theses,
https://doi.org/10.1007/978-981-15-2918-4_7

7.2 Future Recommendations

EMTEs fabricated on the plastic substrates by our low-cost, solution-processed strategies have shown excellent performance in flexible electronic devices. With minor adjustments, we hope these fabrication strategies can also be adapted for fabricating stretchable electronic and biomedical devices such as wearable displays and electronic skins etc.

With superior optical and electrical properties, EMTEs have broad applications in flexible electronic devices. Although the applications presented in this thesis are limited to DSSCs and FTTHs, our EMTEs offer the potential use in other type of flexible solar cells such as perovskite solar cells [1, 2] and other flexible electronic devices that we are currently investigating.

In Chap. 5, nano-EMTEs are fabricated on flexible substrates to verify the dimensional scalability of LEIT and TEIT fabrication strategies. Since nano-EMTEs have narrower metal-mesh linewidths and pitches therefore these electrodes have better electrical continuity and carrier transportation as compared to micro-EMTEs and hence can be more effectively employed in organic electronic devices with short exciton diffusion lengths [3–5].

Due to the relatively larger size of geometric features (pitch, linewidth, and metal-thickness) of the metal-meshes in micro-EMTEs, the optical transmittance and sheet resistance of micro-EMTEs can easily be approximated by the geometric features of the metal-mesh. However unlike micro-EMTEs or other continuous films, the performance of nano–EMTEs (especially with sub-wavelength metal-mesh features) can possibly be influenced by other plasmonic effects in addition to geometric features of the metal-mesh, [6] but further study on this phenomenon and its effect on electrode's performance is needed.

References

1. Tavakoli MM, Tsui K-H, Zhang Q, He J, Yao Y, Li D, Fan Z (2015) Highly Efficient flexible perovskite solar cells with antireflection and self-cleaning nanostructures. ACS Nano 9(10):10287–10295. https://doi.org/10.1021/acsnano.5b04284
2. Yin X, Chen P, Que M, Xing Y, Que W, Niu C, Shao J (2016) Highly Efficient flexible perovskite solar cells using solution-derived NiOx hole contacts. ACS Nano 10(3):3630–3636. https://doi.org/10.1021/acsnano.5b08135
3. Deng B, Hsu P-C, Chen G, Chandrashekar BN, Liao L, Ayitimuda Z, Wu J, Guo Y, Lin L, Zhou Y, Aisijiang M, Xie Q, Cui Y, Liu Z, Peng H (2015) Roll-to-roll encapsulation of metal nanowires between graphene and plastic substrate for high-performance flexible transparent electrodes. Nano Lett 15(6):4206–4213. https://doi.org/10.1021/acs.nanolett.5b01531
4. Gao T, Huang P-S, Lee J-K, Leu PW (2015) Hierarchical metal nanomesh/microgrid structures for high performance transparent electrodes. RSC Adv 5(87):70713–70717. https://doi.org/10.1039/C5RA14851K

5. Gao T, Li Z, Huang P-S, Shenoy GJ, Parobek D, Tan S, Lee J-K, Liu H, Leu PW (2015) Hierarchical Graphene/Metal grid structures for stable. Flexible transparent conductors. ACS Nano 9(5):5440–5446. https://doi.org/10.1021/acsnano.5b01243
6. Muzzillo CP (2017) Metal nano-grids for transparent conduction in solar cells. Sol Energy Mater Sol Cells 169:68–77. https://doi.org/10.1016/j.solmat.2017.04.048

The manufacturer's authorised representative in the EU is Springer Nature Customer Service Centre GmbH, Europaplatz 3, 69115 Heidelberg, Germany. If you have any concerns regarding our products, please contact ProductSafety@springernature.com

Printed and bound by CPI Group (UK) Ltd, Croydon, CR0 4YY
25/09/2026
02222901-0015